GENETICS – RESEARCH AND ISSUES

# BASICS OF MEDICAL MOLECULAR BIOLOGY

# GENETICS – RESEARCH AND ISSUES

Additional books in this series can be found on Nova's website under the Series tab.

Additional E-books in this series can be found on Nova's website under the E-books tab.

GENETICS – RESEARCH AND ISSUES

# BASICS OF MEDICAL MOLECULAR BIOLOGY

TAREK H. EL-METWALLY
GAMIL A. TAWADROUS
YAKOUT A. EL-SENOSI
AND
SAMER M. ZAHRAN

Nova Science Publishers, Inc.
*New York*

For permission to use material from this book please contact us:
Telephone 631-231-7269; Fax 631-231-8175
Web Site: http://www.novapublishers.com



**Library of Congress Cataloging-in-Publication Data**

Basics of medical molecular biology / Tarek H. El-Metwally ... [et al.].
p. ; cm.
Includes bibliographical references and index.
ISBN 978-1-61668-938-4 (hardcover)
1. Molecular biology. I. El-Metwally, Tarek H.
[DNLM: 1. Nucleotides. 2. DNA. 3. Genetic Techniques. 4. Molecular Biology. 5. RNA. QU 57 B311 2010]
QH506.B37 2010
616'.042--dc22

2010022567

*Published by Nova Science Publishers, Inc. ✢ New York*

# Contents

# Preface

Historically, molecular biology was established in the 1930s and the term was first coined by Warren Weaver in 1938 - the director of Natural Sciences for the Rockefeller Foundation. At the time, he believed that biology was about to undergo a molecular revolution as signaled from the significant recent advances in fields such as X-ray crystallography. Molecular biology is a merger between biochemistry and genetics that undertakes the study of the molecular fundamentals of metabolism of the genetic material, i.e., replication, transcription, and translation, and its manipulation for the benefit of life. There is not a defined line between these disciplines. Molecular biology is the molecular three-dimensional structural studying approach of biology as reflected on genesis and function to search below the large-scale manifestations of classical biology. The central dogma of molecular biology, i.e., transcription of the genetic material into RNA and its translation into proteins, provides a simplified but excellent base for comprehending molecular biology, despite the today's more complicated science. It clarifies interactions between various systems of a cell to exchange information, particularly the structure and function of the gene and epigenetics. It utilizes specific techniques along with genetic and biochemistry techniques. Such techniques include; gene and animal cloning, polymerase chain reaction, gel electrophoresis, macromolecular blotting (southern, northern and western), arrays (DNA, RNA, and protein), allele-specific oligonucleotides, Most of the molecular biology investigations are quantitative. The recent merge of molecular biology and computer science developed bioinformatics and computational biology. The study of gene structure and function, i.e., molecular genetics, is amongst the most prominent sub-field of molecular biology. It is worth noting that ~25% of the protein encoding genes is concerned with replication, maintenance and expression of the human genome. This science also overlaps biophysics and population- and phylo-genetics. Understanding the level of information provided in this book paves the road for applying higher and more specialized aspects of molecular biology and molecular genetics. The book highlights the rationale behind most of the related diseases afflicting the nuclear and the mitochondrial genetic systems for specific prevention and/or intervention.

The authors

*Chapter I*

# Structure, Biosynthesis and Degradation of Nucleotides

## Topics Discussed

- Nitrogenous bases, nucleotides and nucleosides and their functions;
- Metabolism of nucleoproteins;
  - Metabolism of Purines;
    - Biosynthesis and Catabolism;
    - Related diseases;
  - Metabolism of Pyrimidines;
    - Biosynthesis and Catabolism; and
    - Related diseases.

## Learning Objectives

After understanding this part, the student should be able to:

- Compare and contrast the structure and biosynthesis of purines and pyrimidines, highlighting the differences between de novo and salvage pathways. And, comprehend how expensive is de novo system to be reflected on malnutrition-poor growth equation.
- Describe how cells meet their requirements for nucleotides at various stages in their cell cycle, at rest and during cell division.
- Describe the metabolic basis and therapy for classic disorders in nucleotide metabolism: Lesch-Nyhan syndrome, gout and severe combined immunodeficiency syndromes (SCIDs).
- Explain the biochemical rationale for using methotrexate, 5-fluorouracil and 5-fluorodeoxyuridine in chemotherapy.
- Describe several metabolic conditions that result in accumulation of orotic acid in urine.

# Nitrogenous Bases and Nucleotides

Nitrogenous bases are the unit heterocyclic compounds participating in RNA and DNA structure and are either purines or pyrimidines. Purines are two: adenine and guanine and their metabolic derivatives, e.g., hypoxanthine, xanthine, and uric acid. Pyrimidines are three: cytosine, uracil and thymine. These bases exist in two forms; keto/amino vs. hydroxy/imino. The keto/amino tautomeric form of these bases is called lactam and occurs normally in the DNA structure, and, the rare enol tautomeric form (hydroxy/imino) form is called lactim. If the lactim form occurs in the DNA structure, it could lead to nucleotide mismatching during DNA polymerization because, e.g., enol-thymine base-pairs with G instead of A. Uracil presents in RNA in place of thymidine that locates to the DNA. The bases, their nucleoside and nucleotide forms, and nomenclature are presented in Table 1.

Repeated structural units of nucleotides connected to each other form nucleic acids which are either DeoxyRibonucleic Acid (DNA) or RiboNucleic Acid (RNA) depending on the sugar component and nature of the bases. Pentoses participating in the free or nucleic acid-incorporated nucleotides are either β-D-ribofuranose (ribose) for RNA or β-D-2-deoxy-ribofuranose (2-deoxyribose) for DNA.

In the nucleus, the basic part of the chromosomes is the DNA that is the storage and expression vehicle for the genetic characters including regulation of protein synthesis and conferring identity. DNA occurs also in the mitochondria with a few genes important for the mitochondrial metabolic functions. RNA is present in the nucleus, cytoplasm, and mitochondria and has a central role in the process of protein synthesis as intermediate between DNA and proteins.

Nucleosides are formed upon the attachment of C1 of the pentose sugar to the N1 of a pyrimidine or N9 of a purine bases. Nucleotides exist in the mono-, di-, or tri-phosphorylated forms. Nucleotide monophosphates are formed by the attachment of a phosphate to -OH group on C5 or C3 of the sugar with a low energy phosphoester bond. If the phosphoric acid is removed from nucleotide, a nucleoside is produced. The connection with the $2^{nd}$ and the $3^{rd}$ phosphate groups is through a high energy phosphodiester bonds (~) - each stores a releasable and reusable 7.3 Kcals/mole. The phosphates are designated α, β and γ, with the α-phosphate being the one directly attached to the sugar.

The nucleoside form of uracil (see the formula below) is called uridine (U) and the monophosphate form is called uridylic acid or uridine monophosphate (UMP) that is always ribose-containing. The ability of nucleotides to absorb the ultraviolet light (UV) explains why UV light causes mutations. This absorpitivity at neutral pH and close to 260 nm also correlates directly with the concentration of nucleic acids in solutions.

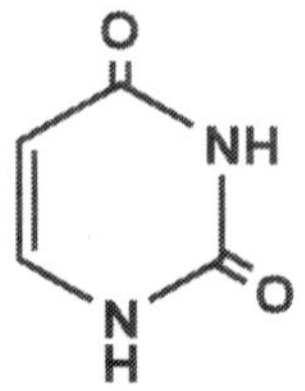

O
NH
O
N
H

**Uracil**

**Table 1. Nucleic acid bases, their nucleoside and nucleotide forms, and, nomenclature**

| Free base form | Nucleoside (Deoxy/Oxy) form | Nucleotide (Deoxy/Oxy) form |
|---|---|---|
| 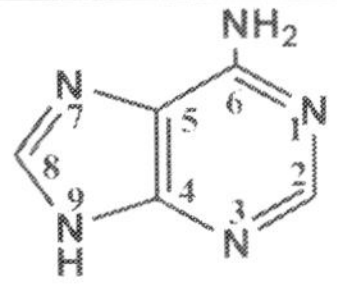 *Adenine* | 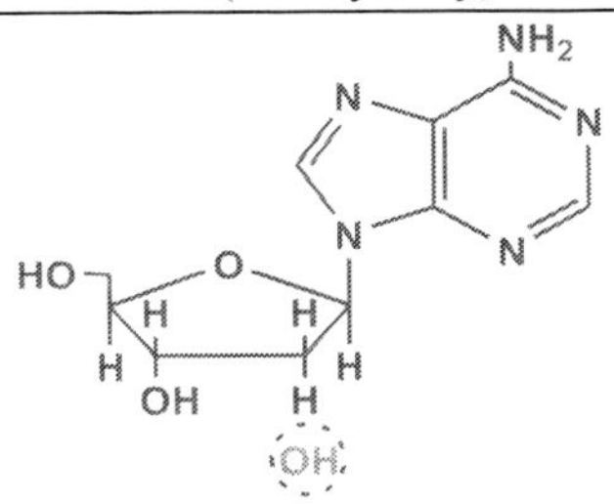 2-Deoxyadenosine, dA (Adenosine, A) | 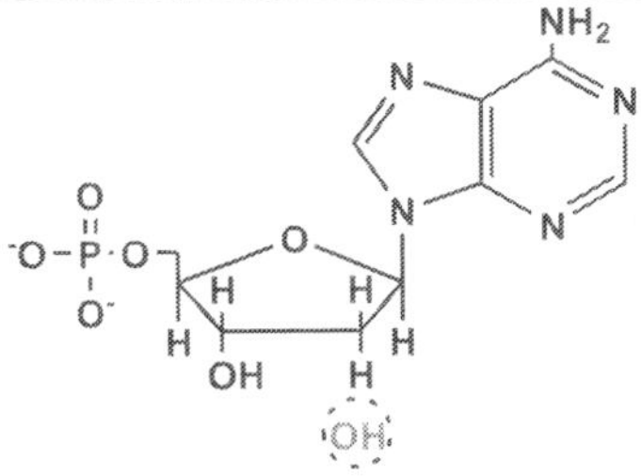 2-Deoxyadenosine monophosphate, dAMP, Deoxyadenylic acid (Adenosine monophosphate, AMP, Adenylic acid) |
| 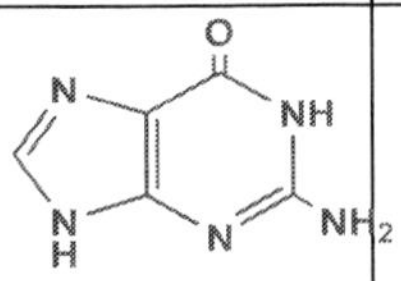 *Guanine* | 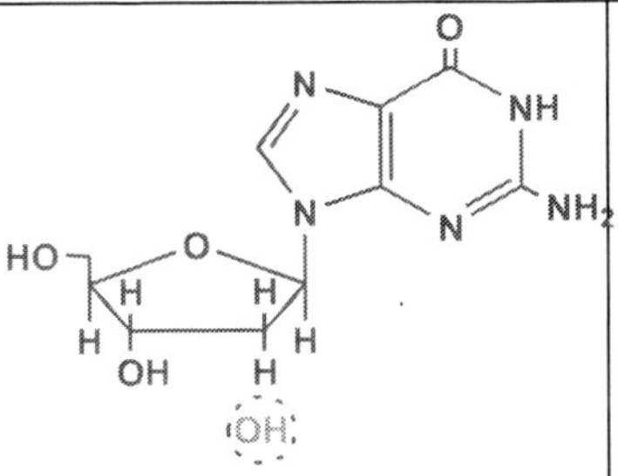 2-Deoxyguanosine, dG (Guanosine, G) | 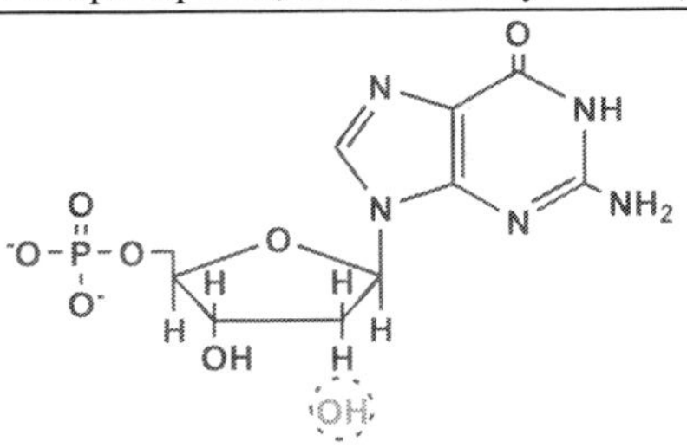 2-Deoxyguanosine monophosphate, dGMP, Deoxyguanylic acid (Guanosine monophosphate, GMP, Guanylic acid) |
| 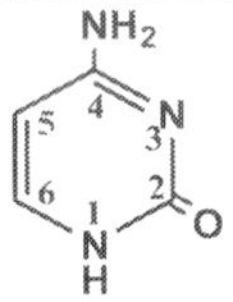 *Cytosine* | 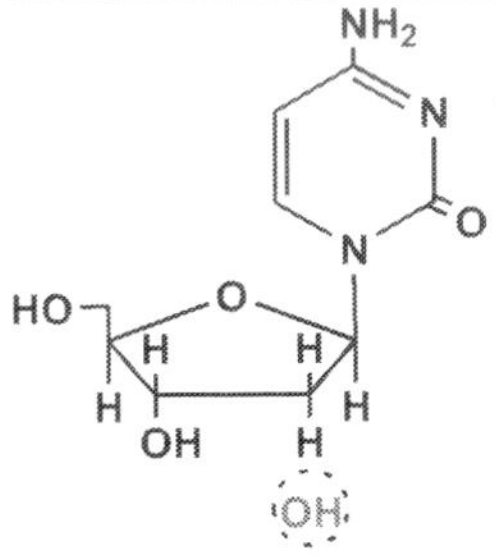 2-Deoxycytidine, dC (Cytidine, C) | 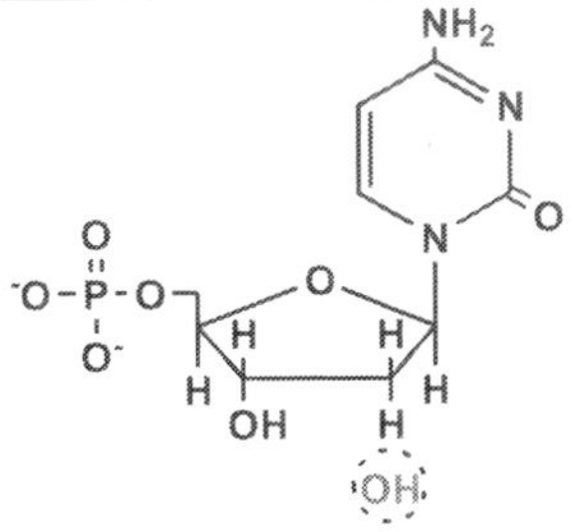 2-Deoxycytidine monophosphate, dCMP, Deoxycytidylic acid (Cytidine monophosphate, CMP, Cytidylic acid) |
| 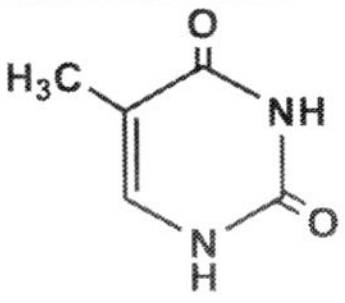 *Thymine* | 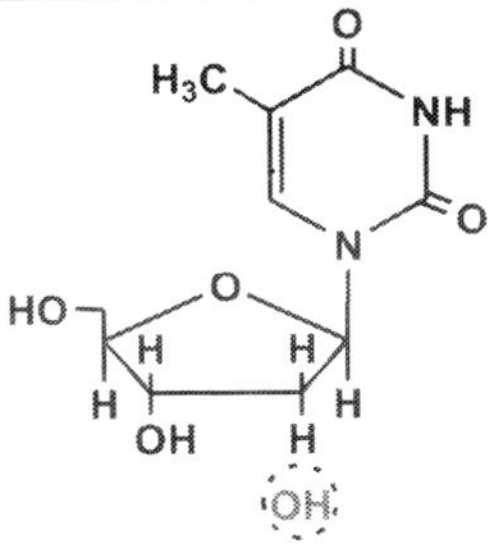 2-Deoxythymidine, dT (Thymidine, T) | 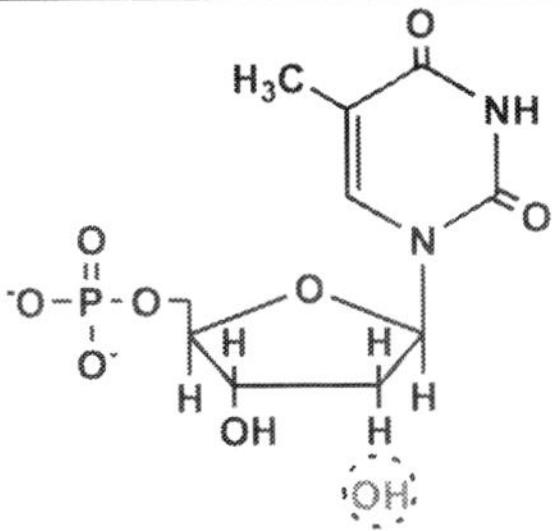 2-Deoxythymidine monophosphate, dTMP, Deoxythymidylic acid (Thymidine monophosphate, TMP, Thymidylic acid) |

*Nucleotides of biological importance:* In addition to forming nucleic acids, a number of nucleotides exist in the free state in living cells and have great biological importance. Examples include:

1. Adenosine, guanosine, cytidine, and uridine triphosphates (ATP, GTP, CTP, and UTP) are used in energy metabolism and activation of intermediary metabolites. ATP/ADP and GTP/GDP are important in storage and utilization of free energy in the living cells.
2. The 3',5'-cyclic-AMP and cyclic-GMP (cAMP and cGMP) serve as second messengers for the action of many hormones.
3. 3-phosphoadenosine-5-phosphosulfate (PAPS) is the active sulfate donor for sulfation reactions.
4. S-adenosyl-methionine is the active form of methionine, which serves as the major methyl donor in transmethylation reactions.
5. Synthetic analogs of normally occurring nucleotides are used in the treatment of cancer, immunosuppression and viral infection. These analogs can act as enzyme inhibitors or falsely replace the natural nucleotides during nucleic acids synthesis. In this way viruses, immune and cancer cell replication are inhibited.
6. Adenosine and ATP function as neurotransmitters in CNS and peripheral nerves.

*Worth Noting: Structure and functions of ATP*

*Structure and functions of the free ATP as an example nucleotide triphosphate* (see, Figure 1)*:* ATP participates in nucleic acid structures; synthesis of dynamic molecules and as a storage form of energy in cells utilized for various metabolic functions. The following is a non exhaustive list of some of the ATP metabolic functions:

1. Muscle contraction and nerve conduction.
2. Absorption, secretion and active transport across membranes through maintenance of electrolyte balance and sodium pump activity.
3. Derive the majority of the anabolic reactions and activates some metabolites, e.g., glucose through kinase activities.
4. Synthesis of cAMP as a major signaling and metabolic regulatory molecule, and active methionine and active sulfate as important methyl and sulfate donors in respective reactions.

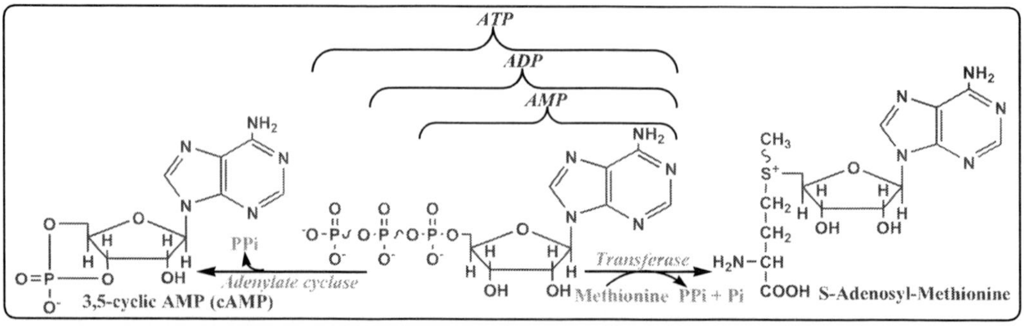

Figure 1. Structure of AMP, ADP, and ATP, and, formation of cAMP and S-Adenosyl-Methionine.

# Nucleoproteins

Nucleoproteins are the nuclear content of any cell; i.e., the functional form of the DNA. The nucleoprotein complex is also called the chromatin that is formed from the cell genome (the DNA content) and the structural and regulatory proteins and RNA of the nucleus.

## Digestion

Plant and animal tissues taken in the diet are formed of cells with their nuclei containing the nucleoprotein. The very small amounts of nucleic acids component of the dietary nucleoproteins is released from the protein component by sequential digestion of the protein by pepsin and trypsin. However, diet differs massively in its nucleic acid content with the highest in the liver to the lowest (almost zero nucleic acid content) in the milk and dairy products and eggs. Plants contain the methyl purines; e.g., caffeine of coffee, theobromine of coca, and theophylline of tea.

The nucleic acid part is hydrolyzed by a number of enzymes of the pancreatic juice [the nucleases: deoxyribonuclease (DNases) and ribonucleases (RNases)] and those of the intestinal juice and mucosa (the polynucleotidases, phosphatases, nucleotidases, and nucleosidases). The hydrolysis products include: nitrogenous bases (purines and pyrimidines), pentoses (ribose or deoxyribose) and inorganic phosphate.

*Nucleoproteins ⇒ nucleic acid + protein*
*Nucleic acid ⇒ Mononucleotides*
*Mononucleotides ⇒ Nucleoside + phosphate*
*Nucleoside + Pi ⇒ nitrogenous base + pentose-1-phosphate*

## Absorption

Ingested purines and pyrimidines are poorly absorbed from the small intestine and fate to be catabolized into uric acid in the liver and intestinal mucosa. Nucleosides are fairly absorbed from the small intestine again to be catabolized or salvaged in the mucosa and liver. However, the parenterally administered (injected) nucleosides/nucleotides are readily incorporated into DNA. This is the experimental basis of using radiolabeled nucleotides, e.g., $^{3}$H-thymidine for labeling and investigating the rate of DNA and RNA synthesis *in vivo* and *in vitro*.

# Metabolism of Nitrogenous Bases

The body requirements of nitrogenous bases are provided by the *de novo* synthesis (i.e., from scratch) from low molecular weight precursors and by the salvage pathway that recycles bases and nucleosides into nucleotides. Thus, dietary requirement of nitrogenous bases is not essential. In some tissues, the salvage pathways are a major source of nucleotides for synthesis of DNA, RNA and coenzyme - as reflected from the inborn errors (genetic diseases) due to mutations affecting enzymes of the salvage pathways. The major control on rate of synthesis and bioavailability of nitrogenous bases, nucleosides and nucleotides is the body needs - controlled by cell cycle effectors and allosteric regulation of the products.

DNA synthesis and its precursor nucleotides are coordinately regulated with the cell cycle progression before and during the S-phase and inhibitors of nucleotide synthesis are utilized as chemotherapy to inhibit cell proliferation in cancer treatment and immunosuppression during organ transplantation. Therefore, nucleotides are not only essential for DNA and RNA synthesis, but also, for energy metabolism (e.g., ATP and GTP), coenzyme formation (e.g., NAD and FAD), synthesis of active intermediates (e.g., S-adenosylmethionine); signal transduction (e.g., cAMP and cGMP); allosteric enzyme regulation (e.g., ATP and cAMP); and in their modified form are used as chemotherapy for cancer, immunosuppression and viral infection.

# Metabolism of Purines

## Biosynthesis

Purine bases include adenine, guanine, hypoxanthine and xanthine. Their biosynthesis takes place in the *liver* mainly. There are two pathways for the synthesis of purine bases:

- The *de novo* pathway; the major pathway for purine nucleotides biosynthesis (Figure 2):

*Site of synthesis:* It takes place in liver and a number of other tissues. Although all body tissues are capable of de novo synthesis including brain and immune cells, the very high expenses of the process makes salvaging bases and nucleosides - synthesized by the liver and secreted into the blood - an easy alternative. This is particularly important for the brain, resting (non-proliferating) and HIV-infected T-lymphocytes.

*Steps:* Transferases activate the formation of nucleoside-5'-phosphates by the stepwise additions onto C1 carbon of 5-phosphoribosyl-1-pyrophosphate (PRPP; the activated sugar intermediate). PRPP synthesis from ribose-5-phosphate in presence of ATP is activated by PRPP synthetase that is *the 1$^{st}$ key regulatory step (Ribose-5-phosphate + ATP ⇒ PRPP + AMP).* This step is synergistically feedback inhibited by accumulation of purine-5'-nucleotides (balanced proportions of GDP and ADP) and is stimulated by availability of ribose-5-phosphate from the pentose phosphate pathway.

Afterwards, there is stepwise addition of; amido group from glutamine with the release of PPi catalyzed by glutamine-PRPP amidotransferase as *the 2$^{nd}$ key regulatory and the committed step*; glycine that requires ATP; formyl-moiety from $N^5N^{10}$-methenyl-tetrahydrofolate ($FH_4$); amido group from glutamine that requires ATP; carboxylation; amino group from aspartate in two steps with release of fumarate catalyzed by adenylosuccinate lyase and ATP; and formyl-moiety from $N^{10}$-formyl-$FH_4$. This is followed by dehydration by cyclohydrolase for ring closure into inosine-5'-monophosphate (IMP), i.e., hypoxanthine-ribose-phosphate or inosine monophosphate. The cascade requires activation by several amidotransferases and transformylases. Isotope precursor studies revealed the specific source of each atom of the purine base - see Figure 2 insert.

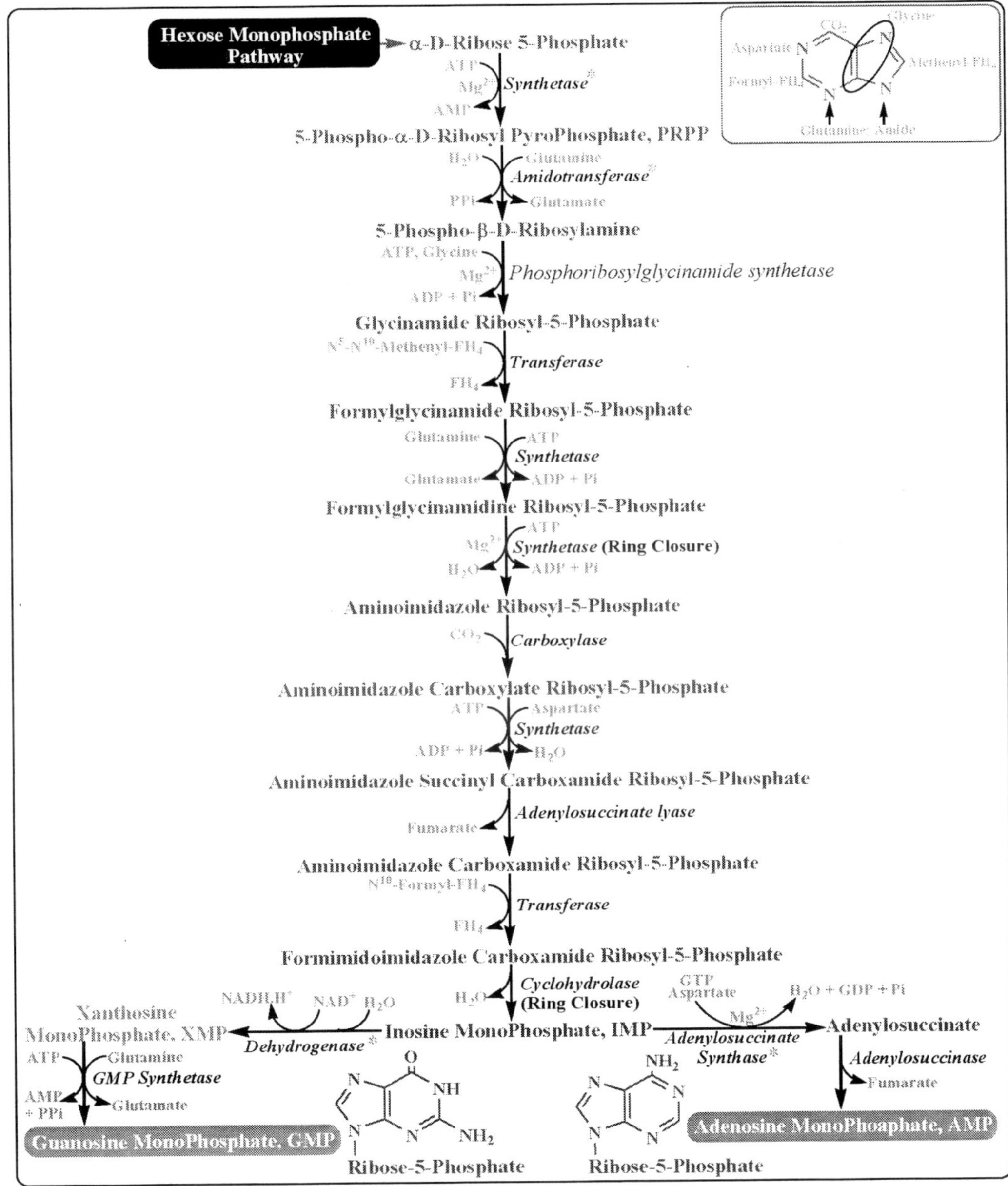


Figure 2. Steps of the *de novo* biosynthesis of purines. The insert illustrates sources of the different atoms of the purine nucleus; (*) denotes key regulatory steps.

After the branch point, addition of the amino group of aspartate onto IMP in two reactions activated by *the 3rd key regulatory adenylosuccinate synthetase* and adenylosuccinase, respectively release fumarate and gives AMP. IMP dehydrogenation into XMP (xanthosine monophosphate) is activated by *the 4th key regulatory IMP dehydrogenase*.

XMP transamidination by glutamine to give GMP is activated by the transamidinase (or GMP synthetase). IMP conversion into AMP is activated by GTP as energy deriver, or into GMP is activated by ATP as energy deriver – through the aforementioned two distinct reaction pathways. Such inter-dependence of AMP on GTP and GMP on ATP provides controlling tool to balance AMP and GMP ratio to near equivalence. The substrate-specific nucleoside monophosphate and the non-specific nucleoside diphosphate kinases convert AMP into ADP and ATP; and, GMP into GDP and GTP, respectively.

## Regulation

There are 4 key regulated reactions; namely, PRPP synthetase, amidotransferase, adenylosuccinate synthase, and IMP dehydrogenase. A primary site of regulation is the synthesis of PRPP. *PRPP synthetase* is simultaneously allosterically inhibited by GDP at one site, and, by ADP at another allosteric site.

Although is key regulatory, PRPP synthetase is not the committed step of purine biosynthesis because PRPP is also used in de novo pyrimidine synthesis and for both the purine and pyrimidine salvage pathways. The committed step of purine synthesis is *glutamine phosphoribosyl amidotransferase* reaction that is also allosterically inhibited by the purine biosynthetic end products; GMP and AMP and their di- and tri-phosphate forms. The inhibition is brought about by dimerizing the monomer active form (133 kDa) of the enzyme upon binding of these end products.

GMP inhibits the activity of *IMP dehydrogenase*, and AMP inhibits *adenylosuccinate synthetase*. There is reciprocal positive regulation in which the synthesis of AMP is dependent on GTP (synthesized from GMP); and, the synthesis of GMP is dependent on ATP (synthesized from AMP) to balance the amounts of AMP vs. GMP. In the same time GMP and AMP are inhibitor allosteric regulators on oneself reaction. This brings about a balanced production of ATP vs. GTP amounts.

Toxic inhibitors of purine synthesis include; azaserine (an experimental carcinogen and a glutamine analogue) and diaza-oxo-norleucine irreversibly inhibits glutamine utilizing enzymes, e.g., amidotransferase and formylglycinamidine synthetase – see formulae below, 6-mercaptopurine inhibits adenylosuccinate and IMP dehydrogenase, and the immunosuppressant mycophenolic acid inhibits IMP dehydrogenase.

$$N^-{=}N^+{=}CH{-}\overset{\overset{\large O}{\|}}{C}{-}CH_2{-}CH_2{-}\underset{\underset{\large NH_2}{|}}{CH}{-}COOH$$

**Diaza-oxo-norleucine**

$$N^-{=}N^+{=}CH{-}\overset{\overset{\large O}{\|}}{C}{-}O{-}CH_2{-}\underset{\underset{\large NH_2}{|}}{CH}{-}COOH$$

**Azaserine**

*Worth Noting: Folate and DNA Synthesis*

The essentiality of folate for DNA (purine and pyrimidine) synthesis, cell division and human growth is inferred from the inhibition of DNA synthesis and cell division upon dihydrofolate reductase inhibition. Dihydrofolate reductase activates folate - the B complex member vitamin - into the tetrahydrofolate ($FH_4$) that is active in the transfer of one carbon moieties anabolic reaction. Therefore, dividing cells are most sensitive to dihydrofolate reductase competitive inhibitors, e.g., methotrexate, aminopterin, and trimethoprim - all are analog of folic acid. These inhibitors are utilized as cancer chemotherapeutic and immunosuppressant for organ transplantation patients.

- *The salvage pathway: the minor pathway for purine nucleotides*

*Salvage pathway recycles purine bases and nucleosides into nucleotide.* It is the sole source for purine nucleotides in some parasites *(e.g., Mycoplasma, Borrelia,* and *Chlamydia).* It is a significant source of purine nucleotides because hypoxanthine accumulation inhibits the de novo synthesis at the 2nd rate limiting amidotransferase step. In mammals, every cell is capable to salvage purines - particularly lymphocytes and brain cells - through the following two mechanisms:

i. *Single Step Direct Conversion* by phosphoribosylation activated by adenosine phosphoribosyltransferase (APRT) giving rise to AMP and by hypoxanthine-guanine phosphoribosyltransferase (HGPRT) that gives either IMP or GMP. Accumulation of AMP inhibits APRT and accumulation of GMP inhibits HGPRT; see the reaction below.

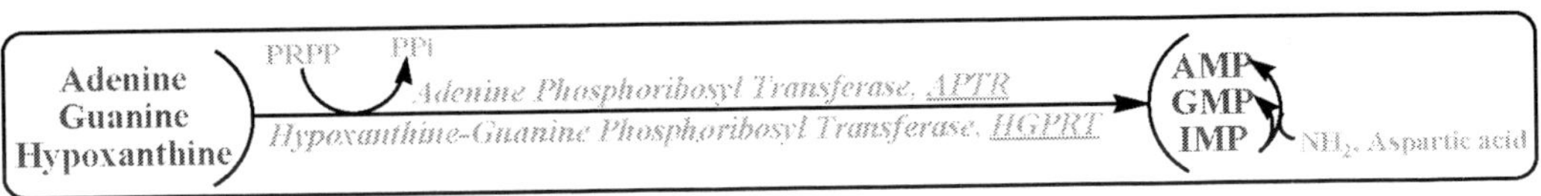


ii. *Two Steps Pathway* reverses the reaction of the purine nucleotide phosphorylases to salvage guanine and hypoxanthine from their catabolic fate. Afterwards, it requires adenosine kinase that converts adenosine and deoxyadenosine into AMP and dAMP. This pathway is less significant compared to the action of phosphoribosyltransferases because mammalian tissues lack the guanosine and inosine kinases. However, deoxycytidine kinase phosphorylates deoxycytidine and deoxyguanosine into dCMP and dGMP; see the reactions below.

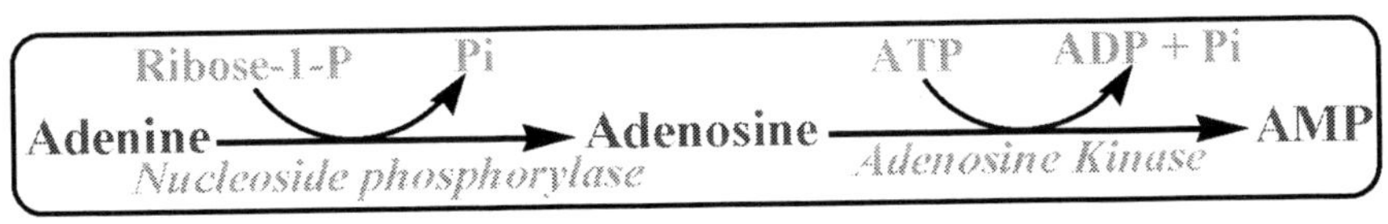


*The purine nucleotide cycle* is a purine salvage process in muscle aiming at generating fumarate that replenishes the Krebs' cycle intermediates particularly in active muscles during large energy demands. It incorporates the *de novo* and salvage mechanisms into a 3-step cyclic AMP-IMP interconversion. *AMP deaminase* is key regulatory in this cycle and its deficiency causes muscle fatigue during exercise; see the reactions below.

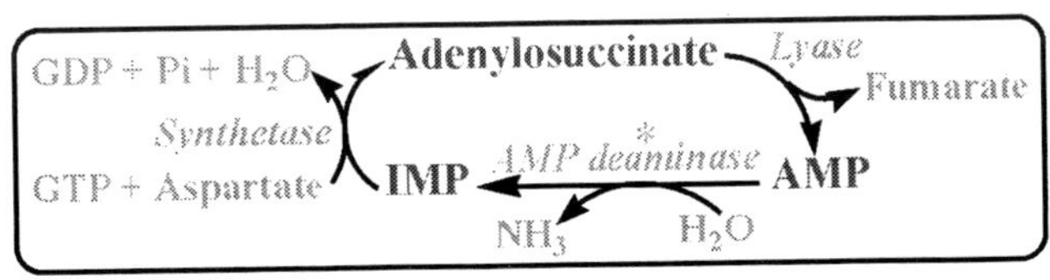

*Deoxyribonucleoside diphosphates:* Ribonucleoside diphosphates (ADP, GDP, CDP and UDP) reduction at C2'-OH of the ribose gives rise to their corresponding 2'-deoxyribonucleoside diphosphates. This is activated by the one enzyme *ribonucleotide reductase complex* in the actively DNA synthesizing cells. The enzyme requires *reduced thioredoxin* that is renewed from the oxidized thioredoxin protein by the *NADPH-thioredoxin reductase*. The enzyme has allosterically regulated binding sites for all nucleotides (oxy- and deoxy-) to modulate the substrate specificity and the overall reaction rate that ensures balanced production of deoxyribonucleotides for synthesis of DNA. There are two major allosteric sites; one controls the overall activity of the enzyme, whereas, the other determines the substrate specificity of the enzyme. ATP binds the overall activity site to activate the enzyme, whereas, dATP binding to this site inhibits the enzyme; detailed as follows. For the substrate specificity, ATP binding to the substrate site activates the reduction of pyrimidines (CDP and UDP) to form dCDP and dUDP. Then dUDP is converted into dTTP. Accumulated dTTP binds to the substrate site and induces the reduction of GDP. Accumulated dGTP displaces dTTP to allow ADP to be reduced to dADP that is phosphorylated into dATP. Accumulated dATP inhibits the overall activity of the enzyme. *The anticancer hydroxyurea inhibits this enzyme.*

## **Catabolism** (Figure 3)

*Site:* Purine catabolism occurs in liver, intestine, spleen, kidney, pancreas, skeletal and heart muscles. In human and higher primates as chimpanzee, the ultimate end product of purine catabolism is uric acid. The starting metabolites are either adenosine or guanosine. The only benefit of such catabolism is clearance of turnover products to avoid their accumulation and feedback inhibitory effect. The process does not generate energy, but intermediates could be salvaged.

*Steps and regulation: Adenosine* is first converted into inosine by adenosine deaminase (a nucleoside deaminase that also uses 2-deoxyadenosine as a substrate). Ribose is then removed from inosine by phosphorylase and inorganic phosphate to form hypoxanthine and ribose-1-phosphate. Hypoxanthine is subsequently oxidized to xanthine and then to uric acid by the dual specific molybdenum-requiring xanthine dehydrogenase (oxidase). *Guanosine* is first converted into guanine by removal of ribose in the form of ribose-1-phosphate by phosphorylase (actually is a nucleosidase). Then guanine is hydrolytically deaminated by guanase to xanthine. Xanthine is then oxidized to uric acid by the xanthine dehydrogenase (oxidase). The molybdenum-sulfide at the active site of the enzyme is where allopurinol inhibitor binds (after being oxidized) to inactivate the enzyme. Purine deoxyribonucleosides are catabolized the same.

*In lower primates and other mammals*, the copper-dependent uricase enzyme converts uric acid to allantoin by oxidative decarboxylation. *In birds and reptiles*, uric acid is the chief end product of not only purines but also protein nitrogen. This allows these animals to excrete a smaller volume of concentrated urine and conserve their body water better than mammals. Because of that, they are called *uricotelic* compared to the *ureotelic* mammals that excrete urea as a major end product for protein nitrogen. Fish are *ammonotellic*, ammonia being the chief end product of protein nitrogen. Methyl purines are catabolized mainly into urea and other products but not to uric acid, therefore, they do not exacerbate gouty hyperuricemic cases.

## Excretion of Uric Acid

About 80% of uric acid formed in the body is excreted in urine. The remaining is probably excreted in the bile. The amount of uric acid excreted in the urine varies between 0.5 - 0.6 gm/day. It is mainly in the form of uric acid because the acidity of urine (~5 pH) prevents sodium urate formation that is much more water soluble than uric acid. Thus, *alkalinization of urine helps preventing uric acid stone formation.* The uric acid excretion is approximately halved on a purine-free diet, showing that about 50% of uric acid excreted in the urine is from the exogenous purine catabolism. Thus, *controlling the dietary supply of purines helps preventing hyperuricemia.*

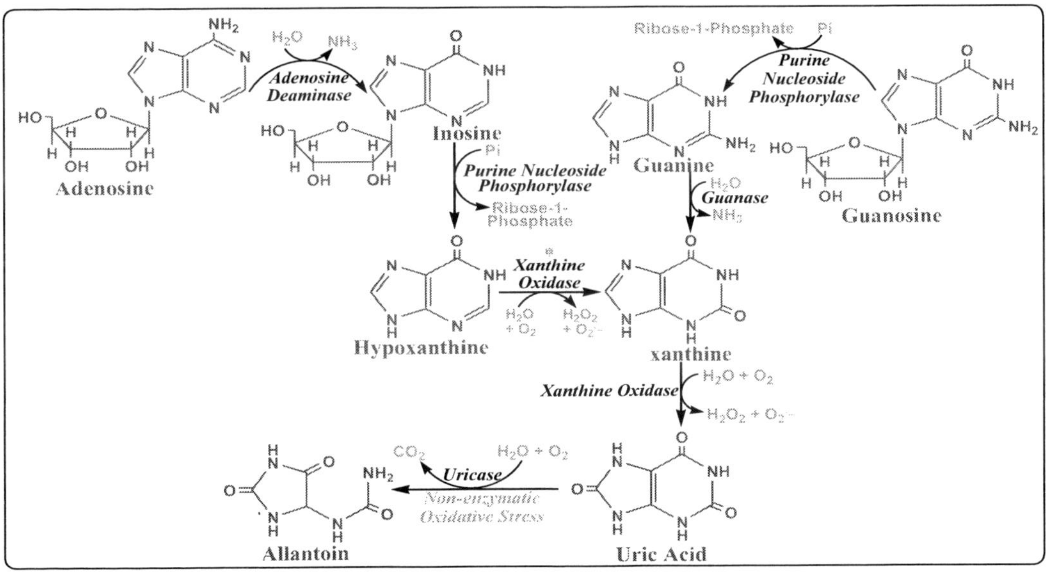

Figure 3. Catabolic steps of purines into uric acid and its enzymatic/non-enzymatic possible oxidation into allantoin. (*) denotes key regulatory.

*Uricolytic Index (UI):* It is an index that shows how much of the uric acid is converted into allantoin in the liver; *UI = allantoin nitrogen/uric acid nitrogen + allantoin nitrogen X 100*

The value of UI is 0 - 2 in primates due to the absence of uricase enzyme. Thus, any allantoin in the urine must be of dietary origin or a product of non-enzymatic oxidation in conditions of oxidative stress. *This is why plasma and urinary allantoin concentration is used as an oxidative stress marker in the human.* The value of UI is 98 - 100 in lower primates and other mammals due to the presence of uricase enzyme, but only 32% in the Dalmatian dog.

*Worth Noting: Xanthine dehydrogenase (Oxidase)*

*Xanthine dehydrogenase (oxidase)* activity is very high in the purine catabolizing tissues, where, it is the rate limiting enzyme, before which purines can be salvaged into nucleotides. It can use $O_2$ as the electron acceptor to give hydrogen peroxide ($H_2O_2$) and superoxide anion ($O_2^{\cdot-}$), or, the safer $NAD^+$ to give NADH. *Allopurinol* (Xyloric – see the formula below) is a suicidal irreversible inhibitor of xanthine oxidase and is used as a drug to decrease uric acid synthesis in cases of gout and as antioxidant in cases with potential oxidative stress such as heart ischemia. Allopurinol gets oxidized into alloxanthine that irreversibly binds and inactivates the enzyme. During ischemia/reperfusion conditions, e.g., heart infarction and tumor with abnormal vascularization, cells are subjected to transient ischemia. Hypoxia induces the excessive hydrolysis of ATP with hypoxanthine accumulation. In normal undamaged cells the predominant form of the enzyme is $NAD^+$-dependent xanthine dehydrogenase (as the safe catabolizing form of xanthine oxidase) that gets proteolytically cleaved during hypoxia and cell damage into the oxidase from. Xanthine oxidase has no damaging effect during hypoxia because of the high *Km* for $O_2$. However, when the tissue is reperfused with blood and the $O_2$ level is normalized, xanthine oxidase works to generate large amount of the damaging oxygen free radicals; namely, $O_2^{\cdot-}$ and $H_2O_2$. *This is why this type of cell damage is called ischemia-reperfusion injury.*

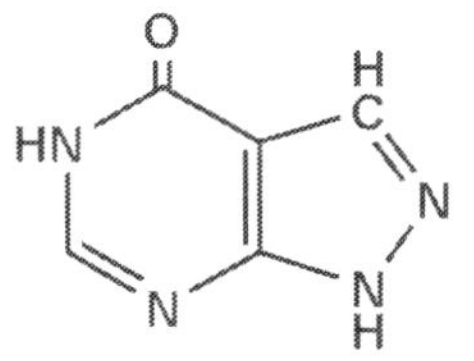

*Allopurinol*

## Diseases due to Defective Purine Metabolism

1. *Lesch-Nyhan syndrome:* Lesch-Nyhan syndrome is a rare *X-linked recessive* disorder that primarily affects the males. It is due to severe deficiency or complete absence of *GHPRT* gene activity located on the X chromosome due to *deletion, substitution or frame shift mutations or due to defective mRNA splicing mutations.*

   Mutation of the *GHPRT* gene causes three main problems that elucidate the importance of salvage pathway particularly for the brain:

   - *The first* is overproduction hyperuricemia. The accumulation of PRPP induces unnecessary *de novo* synthesis of purines that are redirected to the catabolic fate into uric acid. Accumulation of uric acid leads to gout, poor growth and urate stones formation in the kidney and bladder which may be followed by renal failure. Accumulation of PRPP is caused also by mutations of *PRPP synthetase* that lead to high affinity to ribose 5-phosphate and resistance to feedback inhibition with subsequent PRPP overproduction.
   - *The second* problem is self-mutilation at age 2-3. Affected individuals have to be restrained from biting their fingers and tongues and aggression towards the others.

- *Finally*, mental retardation and severe muscle weakness.

*Ex vivo* gene therapy for replacing the gene is in experimental stages. Nowadays, the available medications act to decrease the levels of uric acid.

2. *Adenosine deaminase deficiency:* It is an *autosomal recessive disorder* that leads to one form of *Severe* Combined *Immuno-Deficiency (SCID)* involving both T- and B-cell immunodeficiency characterized by severely reduced number and dysfunction of these cells (bubble boy disease) the patient lacks the thymus gland.

   The insufficiency of the adenosine deaminase enzyme leads to accumulation of DNA catabolic products; dGTP and dATP, that feedback inhibits ribonucleotide diphosphate reductase enzyme and the synthesis of other types of nucleotides. Depletion of DNA precursors causes developmental arrest and apoptosis particularly in tissues with high rate of turnover such as bone marrow and T-cells. Immune cells locating in the blod are particularly prone to this effect because other tissue cells dump these catabolic products into the blood. There is excess excretion of guanine and hypoxanthine nucleosides and hypouricemia.

   Early age bone marrow transplantation is the treatment of choice. Retrovirus-mediated systemic and *ex vivo* gene therapy for replacing the gene is a very promising alternative treatment. The disease was the first to be subjected to trials of gene therapy. The administration of the modified polyethylene glycol-adenosine deaminase conjugate successfully treated the disease through hydrolyzing these catabolic products. Polyethylene glycol conjugation prologs the enzyme's half-life in the blood.

3. *Purine-Nucleoside phosphorylase Deficiency:* It is an *autosomal recessive disorder* that leads to an *immunodeficiency* that affects the T-cells, whereas, B-cell function may be normal. The inhibition of the enzyme has same metabolic consequences like the adenosine deaminase deficiency, though in a milder form.

4. *Adenylate deaminase deficiency* particularly for the muscle-specific isoenzyme of AMP deaminase - myoadenylate deaminase - lead to post-exercise fatigue, cramping and myalgias due to deficiency of Krebs' cycle fumarate intermediate generated in muscles by the purine cycle; see purine cycle.

5. *Xanthine oxidase deficiency:* It is an *autosomal recessive disorder* leading to reduced uric acid *synthesis* and hypouricemia with increased excretion of hypoxanthine and xanthine, thus causing xanthinuria and lithiasis. Hypouricemia is also caused by severe liver damaging diseases.

6. *Gout (hyperuricemia):* It is a metabolic disease characterized by increased levels of uric acid (as *urates*) in the blood (hyperuricemia) above 8.0 mg/dL (normally it is 2.5 - 8 mg/dL) accompanied in most cases with increased excretion of uric acid in the urine.

   Hyperuricemia results in the deposition of needle-shaped crystals of sodium urate in the form of monosodium urate on the joint linings and in soft tissues around it (especially those of the big toe) skin, kidney and other tissues with reduced atmospheric temperature. The precipitated urates are called tophi. Uric acid and urates precipitates in the kidney and urinary tract resulting in uric acid stone formation.

   Urate deposition elicits inflammatory reactions in the joints, i.e., acute gouty arthritis that progresses to a chronic form. Arthritis usually starts in the metatarsophalyngeal joint of the big toe. Gouty tophi produced by phagocytosis of uric acid crystals by polymorphs lead to destruction of these cells with the release of chemotactic substances that attract more polymorphs and exaggerate the acute inflammatory reaction. Gout is associated with morning stiffness that eases on walking opposite to osteoarthritis that is exaggerated by moving.

*Types of gout:*

A) *Primary:* In this case, gout is the primary disease and is either metabolic or renal.

   *Primary metabolic gout* constitutes most of the cases and is due to inherited *autosomal* or *X-linked recessive* metabolic defects characterized by *increased rate of purine synthesis and/or decreased rate of purine salvage* from the breakdown pathway, leading to overproduction of uric acid. These conditions may be accompanied with immunodeficiency. These patients show lowered blood urates (uric acid) levels on a low protein (glycine) diet. One classical example is *Lesch-Nyhan syndrome.* The glycogen storage *Von Gierke's disease* is another example, in which the absence of glucose-6-phosphatase shuttles glucose-6-phosphate to HMP-shunt with excessive production of ribose 5-phosphate, a precursor for PRPP and purine synthesis. This overproduction hyperuricemia with its associating lactemia increases the renal threshold for urate leading to its accumulation in the body.

   *Primary renal gout* is a rare condition and is due to an inherited defect in the kidneys leading to decreased secretion of urates by the renal tubules. Such patients show lowered blood urate levels on a diet poor in nucleic acids, e.g., milk, dairy products and eggs, which are the recommended sources of proteins to treat these cases.

B) *Secondary:* In this case, gout is a complication of other diseases and is either metabolic or renal:

   *Secondary metabolic gout* is due to increased turnover of nucleic acid and uric acid production secondary to increased cell destruction and breakdown of nucleoproteins as in cancer (e.g., leukemia), polycythemia, pneumonia, and psoriasis.

   Secondary *renal gout* is due to decreased excretion of uric acid as in severe renal failure.

*Worth Noting: Biochemical bases for the treatment of gout*

It requires relieve of the pain by using anti-inflammatory drugs; promotion of uric excretion by using uricosuric drugs and alkalinization of the urine; inhibition of uric acid synthesis by competitive xanthine oxidase inhibitors, and dietary control to lower intake of high nucleic acid containing food.

- *Anti-inflammatory agents*, e.g., colchicine, inhibit phagocytosis of uric acid by polymorphs and inhibit the release of chemotactic inflammatory mediators and free radicals in acute gout. Anti-inflammatory uricosuric drugs, e.g., salicylates, cortisone, probenecid and ACTH, act also to increase uric acid excretion by *inhibiting its renal tubular re-absorption.*

- *Alkalinization of urine* increases dissolution of uric acid as sodium urate and prevents its precipitation as stones.
- *The suicidal inhibition of xanthine oxidase* by the structural analogs of hypoxanthine, e.g., allopurinol, prevents formation of uric acid. The accumulated hypoxanthine and xanthine are more readily excreted in the urine since they are more water soluble than uric acid.
- *Dietary control* by avoidance of nucleic acid-rich diets, e.g., liver and meat particularly of old animals and birds is recommended. Dairy products, eggs and meat of young animals are more suitable.

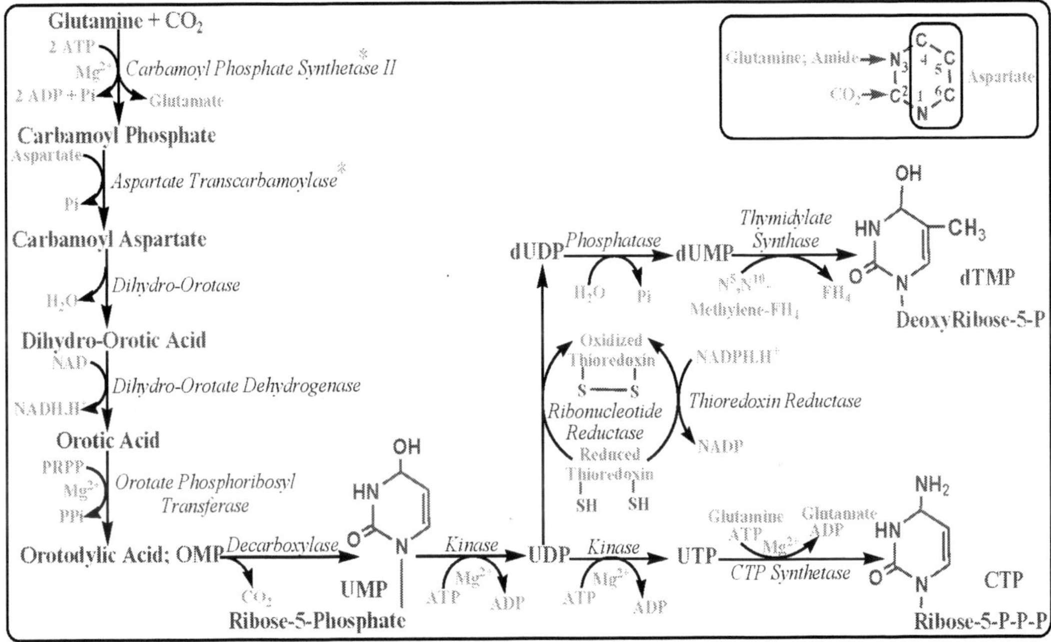


Figure 4. The steps of the de novo biosynthesis of pyrimidines. The insert illustrates the source of the different atoms of the pyrimidine nucleus; (*) denotes the key regulatory enzyme in human and bacteria.

## Metabolism of Pyrimidine

Pyrimidine bases include uracil, cytosine and thymine in nucleic acids. Dihydrouracil, dihydrothymine and orotic acid (6-carboxy-uracil) are intermediates during metabolism of pyrimidines. Unusual or minor pyrimidine bases may occur in nucleic acids, e.g., 5-methylcytosine and 5-hydroxymethylcytosine.

## Biosynthesis

Like purines, pyrimidines are synthesized *de novo* or by salvage pathways. Their site of synthesis is similar to purines.

- *The de novo synthesis pathway (Figure 4):*

The base is first synthesized and then is attached to the sugar (compare to purines). *De novo* synthesis is the major pathway that starts with the cytosolar synthesis of carbamoyl-phosphate by the human key regulatory *carbamoyl-phosphate synthetase II*, a cytoplasmic enzyme that uses glutamine as ammonia donor; does not require N-acetylglutamate as an activator and is inhibited by UTP and activated by PRPP (compared to the mitochondrial carbamoyl-phosphate synthetase I of urea cycle). Carbamoyl phosphate is then added to aspartate activated by *the rate limiting (in bacteria) aspartate transcarbamoylase* step in pyrimidine biosynthesis. The produced carbamoyl aspartate is dehydrated for ring closure and dehydrogenated into orotic acid. In mammals, these three initial steps are executed by one multifunctional polypeptide encoded by a single gene; collectively called CAD (for the first letter of each enzyme activity; carbamoyl phosphate synthetase II, aspartate transcarbamoylase, and dihydroorotase). Orotic acid is ribosylated with PRPP into its mononucleotide form; OMP that is decarboxylated into uracil nucleotide UMP. These two steps are also executed by a multifunctional polypeptide encoded by a single gene; collectively called UMP synthase. The use of multifunctional enzymes at $1^{st}$, $2^{nd}$ and $3^{rd}$ steps and at $5^{th}$ and $6^{th}$ steps ensures efficient channeling of intermediates.

UMP is phosphorylated by kinases into UDP and UTP. UTP is subjected to transamidation from glutamine into CTP activated by CTP synthetase. UDP is the precursors for *de novo* synthesis of the deoxythymine nucleotide; dTMP, through; reduction in presence of reduced thioredoxin, dephosphorylation and transmethylation (from $N^5$,$N^{10}$-methylene-$FH_4$) activated by thymidylate synthase. Radioisotope usage revealed sources of different atoms in pyrimidine structure as indicated in Figure 4 insert.

Like purines, conversion of ribonucleotide diphosphates into corresponding 2-deoxy forms is activated by *ribonucleotide reductase* requiring reduced thioredoxin as hydrogen donor that is regenerated on expense of $NADPH.H^+$ activated by thioredoxin reductase.

- *The salvage pathway:*

As in the two-step purine salvage, it is the minor pathway that uses kinases to phosphorylate and recycle the pyrimidine nucleoside into nucleotides. Each of thymine preferentially reacts with deoxyribose-1-phosphate and uracil preferentially reacts with ribose-1-phosphate to give thymidine or uridine and Pi catalyzed by the highly specific thymine phosphorylase or the relatively non-specific uridine phosphorylase, respectively. Cytidine or uridine is phosphorylated on the expense of ATP into CMP or UMP activated by uridine-cytidine kinase. Deoxythymidine or deoxycytidine is phosphorylated on the expense of ATP into dTMP or dCMP activated by thymidine or deoxyuridine kinase, respectively. Further phosphorylation into diphosphate and triphosphate nucleosides is activated by specific kinases - most important is the *thymidylate kinase;* see the reactions below.

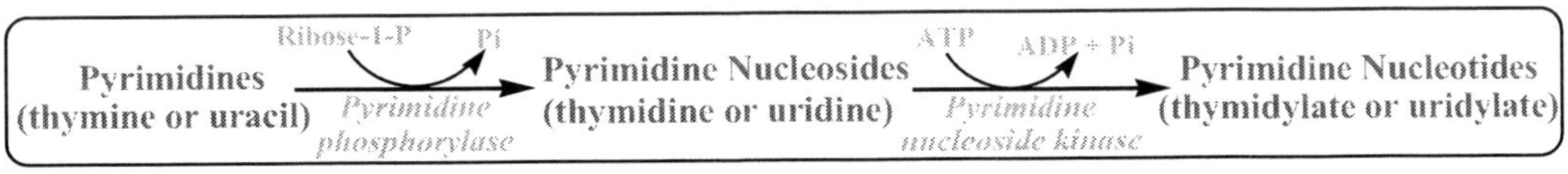

Pyrimidines (thymine or uracil) —[Ribose-1-P → Pi; *Pyrimidine phosphorylase*]→ Pyrimidine Nucleosides (thymidine or uridine) —[ATP → ADP + Pi; *Pyrimidine nucleoside kinase*]→ Pyrimidine Nucleotides (thymidylate or uridylate)

## Regulation of Pyrimidine Synthesis

The synthesis is largely controlled through product feedback allosteric regulation. The key regulatory *carbamoyl phosphate synthetase II* in humans is inhibited by UTP (and by purine nucleotides) and activated by PRPP. The sensitivity to activating effect of PRPP is maximized close to S-phase of the cell cycle, where sensitivity to the inhibitory effect of UTP is minimized; i.e., the enzyme is more readily activated. The opposite happens at the end of the S-phase. This alteration in the sensitivity of the enzyme is due to its activating phosphorylation by the Mitogen *Activated Protein Kinase (MAPK)* in cells committed to cell division. But phosphorylation at a different site by the cAMP-dependent protein kinase makes carbamoyl phosphate synthetase II more readily inhibitable. CTP inhibits and ATP activates *aspartate transcarbamoylase* leading to a strict balance between thymine and cytosine nucleotides. This also reflects coordinate regulation of purine and pyrimidine biosynthesis mole for mole and accumulation of both type of nucleotides allosterically inhibit PRPP synthase, a key regulatory step for both. Coordinated repression and derepression control the first 3 steps into dihydroorotic acid and $5^{th}$ and $6^{th}$ steps into UMP. The anticancer 6-azauridine and allopurinol as alternative substrates inhibit orotidylate decarboxylase.

*Worth Noting: Thymidylate synthase*

Among the many important enzymes working in the de novo synthesis of pyrimidines is *thymidylate synthase*. Thymidine synthase is allosterically inhibited by dTTP and its activity in a given cell is closely related to the cell proliferation rate; particularly at the beginning of the S phase. Radiolabeled thymidine is widely used for isotopic labeling of DNA for radioautographic detection of DNA and/or determining the rate of DNA synthesis. Moreover, its key regulatory importance is invested to inhibit pyrimidine synthesis by modified pyrimidines as irreversible suicidal substrates, such as *5-fluorouracil* (see the formula below) and *5-fluorodeoxyuridine*. Their inhibitory effect on DNA synthesis is used as an *anticancer therapy* approach. Because thymidine is DNA-specific, inhibition of thymidine synthesis via thymidylate synthase using thymidine analogs preferentially inhibits DNA synthesis and not RNA synthesis - as the case with purine analogs.

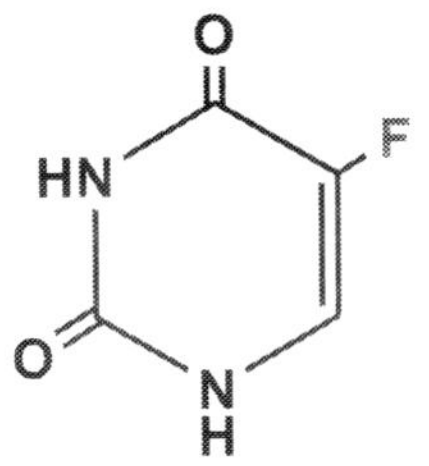

O
HN F
O N
H

**5-Fluorouracil**

## Catabolism

Pyrimidine nucleotides are first dephosphorylated by 5'-nucleotidase into nucleosides. The nucleosides are de-ribosylated by phosphorolysis into pyrimidines. The free bases are catabolized into $NH_3$, $CO_2$, β-alanine and β-aminoisobutyrate; see Figure 5. $NH_3$ is converted into urea. β-Aminoisobutyrate is normally transaminated into methylmalonate semialdehyde that gives succinyl-CoA. The β-amino-isobutyrate is excreted in urine in large amounts (*β-amino-isobutyrate aciduria*) in cases with *rapid DNA turnover*, e.g., leukemias and after exposure to X-rays and other ionizing radiations. β-Alanine is oxidatively deaminated and decarboxylated into acetate. The process does not generate energy, and other than recycling by the salvage pathway, the only benefit of such catabolism is clearance of turnover products to avoid their feedback inhibitory effect.

Pseudouridine as an unusual nucleoside produced from RNA turnover goes unchanged into urine (as the name imply) due to absence of a specific hydrolase or phosphorylase.

## Diseases due to Defective Pyrimidine Metabolism

Most of the defects in the catabolism of pyrimidines do not cause clinical diseases because of the water solubility of their metabolic intermediate. In cases of increased purine breakdown and hyperuricemia, there is also an increased pyrimidine breakdown.

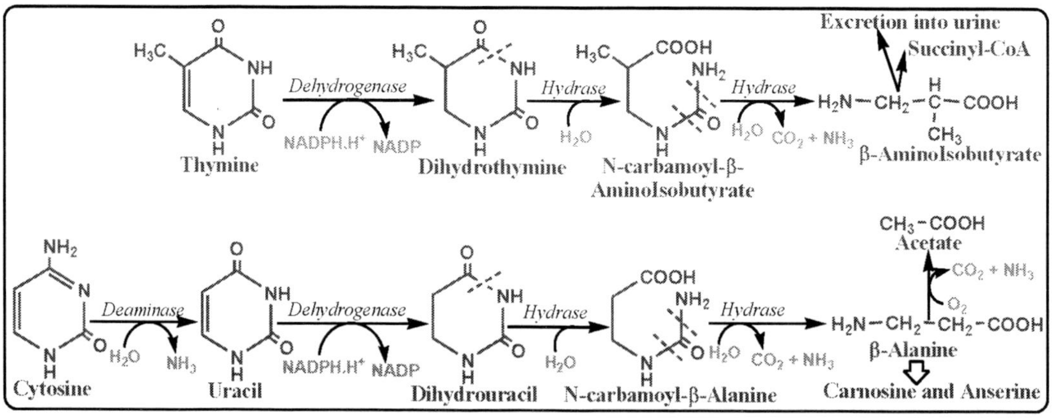

Figure 5. The catabolic fate of pyrimidines; thymine and cytosine.

Orotic acidosis and orotic aciduria types I and II are very rare genetic condition accompanied with poor growth and hypochromic megaloplastic anemia with defective maturation of red cells and leukopenia. They are treatable by supplementing cytidine or uridine that replenishes the deficiency by being converted into UMP. This also feedback inhibits carbamoyl phosphate synthetase II and orotic acid synthesis to ease the acidosis. Allopurinol and azauridine cause orotic aciduria because their catabolites inhibit OMP decarboxylase.

Orotic acidosis and acidurias are due to:

1. *Reye's syndrome* that is characterized by abnormal and severely damaged mitochondria that leak carbamoyl-phosphate of urea cycle into cytosol where it is utilized for excessive synthesis of pyrimidine precursors leading to orotic aciduria.
2. *Type I orotic aciduria* that is due to deficiency of enzymes converting orotic acid into UMP during pyrimidine synthesis (i.e., phosphoribosyl transferase and decarboxylase) is the most severe with accumulation of orotic acid and its excessive excretion in urine with possible precipitation as crystals. The rare *type II orotic aciduria* is due to deficiency of the later enzyme only and is milder.
3. Decreased activity of liver ornithine transcarbamoylase and argininosuccinic acid synthetase of the urea cycle leading to accumulation of carbamoyl phosphate and its osmotic leakage into cytoplasm to be utilized for excessive synthesis of pyrimidine precursors leading to *mild orotic aciduria* and increased uracil and uridine excretion. It may cause hepatic encephalopathy.

## Review Questions

A. *Write short note on:*
   1. Key regulatory reactions of the de novo biosynthetic pathway of purines.
   2. Biochemical bases of treating gout.
   3. Lesch-Nyhan syndrome.
   4. Severe Combined Immuno-Deficiency (SCID).
   5. Orotic acidurias.

B. *Explain the following:*
   1. Overflow of the urea cycle can cause orotic aciduria.
   2. Thymidylate synthase is a target for anticancer chemotherapy.
   3. Although is a minor synthetic pathway, purine salvage pathway could be implicated is harsh inherited diseases.
   4. The glycogen storage von Gierke's disease could be a cause of hyperuricemia.
   5. The role of free nucleotides in biochemistry is as essential as their polymerized role as nucleic acids.
   6. Folate antimetabolite is used as anticancer chemotherapy.

C. *True and False Questions:*
   1. Opposite to the ingested nucleotides, parenterally administered (injected) nucleotides are readily incorporated into DNA and RNA.
   2. Oral but not parenterally injected radio-labeled nucleotides are used to label newly synthesized DNA and RNA.
   3. DNA synthesis and cell division are inhibited upon dihydrofolate reductase inhibition by, e.g., methotrexate, aminopterin, and trimethoprim.
   4. Dividing cells are most sensitive to dihydrofolate reductase competitive inhibitors that are used as cancer chemotherapeutics and immunosuppressants.
   5. The salvage purine synthesis is particularly functional in lymphocytes.
   6. The salvage nucleotide synthesis is dispensable in all of the human body cells.

7. The purine nucleotide cycle is a salvage process in muscle that generates fumarate as an intermediate in the Krebs' cycle.
8. AMP deaminase is key regulatory in this cycle and its deficiency causes muscle fatigue during exercise.
9. The human normal Uricolytic Index is 0 - 2 due to the absence of uricase enzyme.
10. In human, allantoin is a normal purine catabolic end product.
11. The X-linked recessive Lesch-Nyhan syndrome reflects the importance of purine salvage pathways particularly for the brain.
12. Lesch-Nyhan syndrome is due to severe deficiency or complete absence of GHPRT gene activity.
13. The severe combined immuno-deficiency is due to the deficiency of adenosine deaminase activity.
14. The deficiency of the purine-nucleoside phosphorylase causes a milder immunodeficiency than adenosine deaminase deficiency.
15. Deficiency of the muscle-specific adenylate deaminase isoenzyme causes post-exercise fatigue, cramping and myalgias.
16. In the purine cycle, adenylate deaminase acts to generate fumarate for Krebs' cycle.
17. The anti-gout allopurinol (Xyloric) is a suicidal irreversible inhibitor of xanthine oxidase.
18. Avoidance of nucleic acid-rich diets is not essential for controlling gout since urinary uric acid is from endogenous purine.
19. Orotic acidosis and aciduria is due to accumulation of carbamoyl-phosphate in the mitochondria.
20. The synthesis of purines and pyrimidines is coordinately regulated with the cell cycle.
21. Modified pyrimidines are irreversible suicidal inhibitors of thymidylate synthase.
22. Inhibition of thymidylate synthase by 5-fluorouracil and 5-fluorodeoxyuridine is used as an anticancer therapy approach.

D. *Supply the Missing Information questions:*

1. Hyperuricemia ensue when blood uric acid level exceeds ---------------.

   Because human is devoid of ---------------, allantoin is not a normal purine catabolic end-product.

   Hydroxyurea is an antimetabolite anticancer chemotherapy because it inhibits the ----------------------------------- activity.

E. *Multiple Choice Questions:*

1. Reduction in activity of one of the following enzymes causes gout:
   A. Glutamine phosphoribosyl amidotransferase.
   B. Glucose 6-phosphatase.
   C. Glucose 6-phosphate dehydrogenase.
   E. Purine nucleoside phosphorylase.
2. Allopurinol as a treatment for gout works by inhibiting conversion of:
   A. AMP to XMP.
   B. Xanthine to uric acid.
   C. Inosine to hypoxanthine.

D. IMP to XMP.
E. Adenosine to inosine.

3. The chief end product of purine catabolism in man is:
   A . Urea.
   B. Uric acid.
   C. Ammonia.
   D. Allantoin.
   E. Creatinine.
4. The most abundant intracellular free nucleotide:
   A. UTP.
   B. $FAD^+$.
   C. $NADP^+$.
   D. $NAD^+$.
   E. ATP.
5. The normal level of uric acid in the human blood is:
   A. 2.5 – 8 mg/dL.
   B. 80 –110 mg/dL.
   C. 20 – 40 μg/dL.
   D. 150 – 250 mg/dL.
   E. None of the above.
6. Lesch-Nyhan syndrome is:
   A. X-linked dominant GHPRT deficiency.
   B. X-linked dominant APRT deficiency.
   C. X-linked recessive APRT deficiency.
   D. X-linked recessive GHPRT deficiency.
   E. Autosomal dominant GHPRT deficiency.
7. Lesch-Nyhan syndrome is characterized by which of the following characteristics:
   A. Uric acid lithiasis.
   B. Self mutilation.
   C. High intracellular PRPP.
   D. All of the above.
   E. None of the following.
8. All of the following cause hyperuricemia EXCEPT:
   A. Lesch-Nyhan syndrome.
   B. Cancer.
   C. Adenosine deaminase deficiency.
   D. Von-Gierke's Disease.
   E. Severe renal failure.
9. Its deficiency causes immunodeficiency disease:
   A. Hypoxanthine-Guanine phosphoribosyl transferase.
   B. Xanthine oxidase.
   C. PRPP synthetase.
   D. Adenosine deaminase.
   E. HGPRT.

10. Inhibition of ----------- is a target for anticancer chemotherapy:
    A. Dihydrofolate reductase.
    B. Xanthine oxidase.
    C. Thymidylate synthase.
    D. Adenosine deaminase.
    E. A + C.
    F. B + D.
11. The disorder in which the patients have an irresistible urge to bit their finger and lips.
    A. Gout.
    B. Severe combined immunodeficiency disease.
    C. Lesch-Nyhan syndrome.
    D. Reye's syndrome.
    E. Von Gierke's disease.
12. A vitamin that is essential for biosynthesis of purine nucleotides:
    A. Folic Acid.
    B. Niacin.
    C. Cobalamin.
    D. Pyridoxal-amine.
    E. Vitamin K.
13. Deoxyribonucleotides are formed from ribonucleotides by a reduction process catalyzed by:
    A. Ribonucleotide reductase.
    B. Ribonucleotide oxidase.
    C. Ribonucleotide synthase.
    D. Ribonucleotide dehydrogenase.
    E. Ribonucleotide lyase.
14. Carbamoyl phosphate synthetase I and II are similar for:
    A.Carbon source.
    B. Intracellular location.
    C. Nitrogen source.
    D. Regulation by N-acetyl glutamate.
    E. Subcellular location.
15. Lesch-Nyhan syndrome has an enzyme deficiency that catalyzes conversion of:
    A. Adenine to AMP.
    B. Adenosine to AMP.
    C.Guanine to GMP.
    D. Guanosine to GMP.
    E. GDP to GTP.
16. Considering carbamoyl-phosphate synthetase II - all are CORRECT except:
    A. Its product is used for urea synthesis.
    B. It is a cytosolar enzyme.
    C. It uses glutamine as the donor of ammonia.
    D. It is a cytosolar enzyme.

17. Considering carbamoyl-phosphate synthetase II; all of the following are CORRECT except:
    A. It functions in the cytoplasm.
    B. It is the key regulatory enzyme in the de novo pyrimidine biosynthesis.
    C. It is regulated by the availability of ammonia.
    D. It does not require N-acetyl-glutamate as an activator.
18. Considering xanthine oxidase; all are CORRECT except:
    A. Its inhibition by allopurinol is an antioxidant measure.
    B. It converts allopurinol into alloxanthine that irreversiblely inactivates the enzyme.
    C. The NAD-dependent for of the enzyme generates oxygen free radicals.
    D. The high Km for $O_2$ makes the enzyme particularly damaging during reperfusion.
19. Considering xanthine oxidase; all are CORRECT except:
    A. It functions normally as a dehydrogenase with NAD as hydrogen acceptor.
    B. In ischemic tissues it is proteolytically converted into an oxidase.
    C. The oxidase function has a very low $K_m$ for $O_2$.
    D. It is a major source of intracellular oxygen free radicals.

*Answer key for True/False:* 1, T; 2, F; 3, T; 4, T; 5, T; 6, F; 7, T; 8, T; 9, T; 10, F; 11, T; 12, T; 13, T; 14, T; 15, T; 16, T; 17, T; 18, F; 19, F; 20, T; 21, t; 22, T.

*Answer key for Supply the missing information:* 1, 8.0 mg/dL; 2, uricase; 2, ribonucleotide reductase.

*Answer key for MCQs:* 1, B; 2, B, 3, B; 4, E; 5, A; 6, D; 7, D; 8, C; 9, D; 10, E; 11, C; 12, A; 13, A; 14, A; 15, C; 16, A; 17, C; 18, C; 19, C.

*Chapter II*

# The Deoxyribonucleic Acid (DNA): From Structure to Function/Malfunction

## Topics Discussed

- DNA Structure and Types;
- Chromatin Organization;
- Types of DNA sequences;
- DNA replication (pro- vs. eukaryotic), regulation and inhibition;
- Telomerase, telomere and related diseases;
- Cell cycle regulation and abnormalities;
- DNA mutation, repair and related diseases;
- Chromosomal abnormalities and basic cytogenetic and fluorescent *in situ* hybridization techniques;
- Mitochondrial DNA and related diseases.

## Learning Objectives

After reading this part the student should be able to:

- Describe the composition and structure of DNA, based on the Watson-Crick's Model, including the concepts of directionality and complementarily in DNA structure.
- Outline the process for packaging of DNA in the nucleus through different patterns of chromatin structure with a elaborate description of metaphase chromosome.
- Explain how replication of DNA is achieved with high fidelity in a bidirectional manner and in a semi-conservative fashion.
- Discuss the enzymes involved, the activities at replication forks, and the structures and intermediates participating in the replication process.

- Describe the mechanism by which replication is controlled in the eukaryotic cell.
- Explain the mechanism of action of Zidovudine for treatment of HIV-AIDS as an inhibitor of retroviral reverse transcriptase
- Describe the nature of the damage and mechanisms involved in the repair of damage to DNA.
- Explain the biochemical bases of mitochondrial diseases and their pattern of inheritance.

## Nucleic Acid Being the Genetic Material

A few pioneering experiments in the first half of the last century pointed to DNA as the genetic material rather than proteins as it was initially thought. The *Streptococcus pneumoniae* transformation experiment of Frederick Griffith showed that a mixture of the averulant un-encapsulated nonpathogenic strain with heat-inactivated virulent encapsulated strain of the bacterium caused the disease in mice, and, his was subsequently able to isolation of the virulent strain from dead mice. The nature of the transforming (averulant un-encapsulated to virulent encapsulated) principle was confirmed later through Oswald Avery and colleagues to be DNA using biophysical tests and pre-digestion with DNases vs. hydrolytic enzymes for other macromolecules. The strongly suggestive second experiment of Hershey and Chase used radio-labeled bacteriophages - made of the core DNA and capsid protein. The T2 phage was labeled for DNA using the radioactive phosphorus isotope, $^{32}P$, and, for protein using the radioactive sulfur isotope, $^{35}S$ because they wanted to know which component is infective. Incubation of the lytic T2 phage with *Escherichia coli* for a few minutes and agitating the culture in a blender to detach the viral capsid form the bacterial wall followed by low speed centrifugation pelleted the intact bacterial cells that contain the infective viral material from the fluid that contain the remaining viral parts. They found that 70% of the phosphorus radioactivity was in the pellet, whereas, only 20% of the sulfur radioactivity was in the pellet. This indicated that the DNA is most probably the infective principle. Incubating the bacteria with the virus for 20 minute at 37 °C allowed a complete lytic cycle and the release of new virus particles. These new phages contained half of their DNA radio-labeled but only 1% of their protein radio-labeled; possibly a contaminant. Therefore, the genetic material transferred to the new generation of phage was DNA.

## Structure of the Nucleic Acid

Nucleic acids are high molecular weight molecules formed of thousands of building units called nucleotides. Each nucleotide is formed of a nitrogenous base, a pentose and a phosphoric acid molecule. On the basis of the sugar moiety they contain, nucleic acids can be of two types, namely, deoxyribonucleic acid (DNA) and ribonucleic acid (RNA). Nitrogenous bases are connected to carbon one (C1) of the pentose and phosphoric acid is connected to carbon five (C5) of the pentose. Other than the regular bases, naturally occurring modified bases in DNA and RNA include; inosine due to deamination of a preformed adenine, thymine with D-ribose due to transmethylation of a preformed UMP, 5-methylcytosine, 5-hydroxymethylcytosine, mono- and di-N-methylated adenine and 7-methylguanine due to their transmethylation on the expense of S-adenosylmethionine. They regulate sequence recognition and stability.

In the polynucleotide form, the nucleotides are interconnected by phosphodiester bonds most commonly formed between -OH of C3 of the pentose of a nucleotide with -OH of the phosphoric acid attached to C5 of the pentose of the subsequent nucleotide with the release of $H_2O$. This direction of polymerization makes the first nucleotide at the left side with a free ribose-C5 carbon having a phosphate group; an end that is called because of that *the 5'-end*. Polymerization continues to the last nucleotide to the right side of the polymer that will have a free ribose-C3-OH; an end that is called because of that *the 3'-end*. Therefore, the primary structure of DNA and RNA stands exhibit directionality or polarity from left 5' end towards the right 3' end. Their sequence is written/read in the same 5'⇒3' direction, e.g., *5'-pTpCpCpGpApTpCpA-3'*. Each phosphodiester bridge has a negative charge. This negative charge repels nucleophilic species such as hydroxide ion; consequently, phosphodiester linkages are much less susceptible to hydrolytic attack than are other esters such as carboxylic acid esters. This resistance is crucial for maintaining the integrity of information stored in nucleic acids. The absence of the -OH on C2 of deoxyribose of DNA further increases its resistance to hydrolysis by, e.g., alkali - a major distinction from RNA that is alkali liable due to its C2 -OH.

The genomic DNA sequences frequently have complementary sequences that are readable the same in opposite directions and are accordingly called *palindromes* or *palindromic sequences* (palindrome = inverted phrase), e.g., 3'-ACTAGT-5'/5'-TGATCA-3'. They have several biological importance including; most of the target sequence for the DNA cutting enzymes (restriction enzymes) are palindromic sequences; serve as specific DNA binding sites, where transcription regulatory proteins bind to control rate of gene expression; and, have a strong influence on the three dimensional (3D) structure of the DNA molecules. In the RNA, palindrome results in a perfect hairpin, with no loop. Such effect on the RNA 3D structure reflects the biological importance of the palindromes in RNA. In RNA, their biological function is also derived from their ability to bind complementarily with specific gene sequences and affect gene rate of expression, e.g., the inhibitory small nuclear RNAs (see RNA later).

The genetic code sequence of a DNA polynucleotide is read in 3'⇒5' direction in DNA template during replication and transcription. However, the genetic code of an mRNA is read in 5'⇒3' direction during translation. Because DNA is double-stranded, the length of a DNA molecule is expressed in the simplest length unit of one base pairs (bp); one kilobase pair (kb, $10^3$ bp); one megabase pair (Mb, $10^6$ bp); or one gigabase pair (Gb, $10^9$ bp).

Actual life form could be defined as the DNA and the organisms are merely vehicles that the DNA makes to facilitate replication. Thus organisms are the phenotype encoded by the DNA to assure transmission of the genetic information. DNA is the stored form of genetic information encoding genes and carries out, regulates and controls the protein synthesis process. On the other hand, proteins express the genetic characters and maintain and propagate DNA. Therefore, DNA executes genetic control and identity through proteins.

A message copy of the DNA part concerned must be created to be transported to the cytoplasm where it controls and directs protein synthesis, i.e., mRNA. Ribosomes - ribonucleoprotein complexes containing rRNA - are the sites and machinery for of protein synthesis; mRNA is the assembly blue print, and tRNA is the carrier for the protein building units (amino acids).

# Structure of DNA

*Primary structure of DNA* is the linear sequence of its building nucleotide units connected by the phosphodiester bonds. DNA is present in the form of double stranded molecule that acquires it a *helical secondary structure*. Each strand is a very long molecule containing thousands of nucleotides. The two strands bind one another in an anti-parallel, i.e., one strand runs in 5'⇒3' direction (i.e., the coding or sense strand) while the other strand runs in 3'⇒5' direction (i.e., the template, antisense, or non-coding strand). The two strands are complementary to each other and the ratio of purines (adenine + guanine) to pyrimidines (thymine + cytosine) is ~ 1. At molar amount adenine equals thymine and guanine equals cytosine. This rule infers a constant ratio of purine and pyrimidines was coined by Ervin Chargaff and is known as Chargaff's rule (Figure 6).

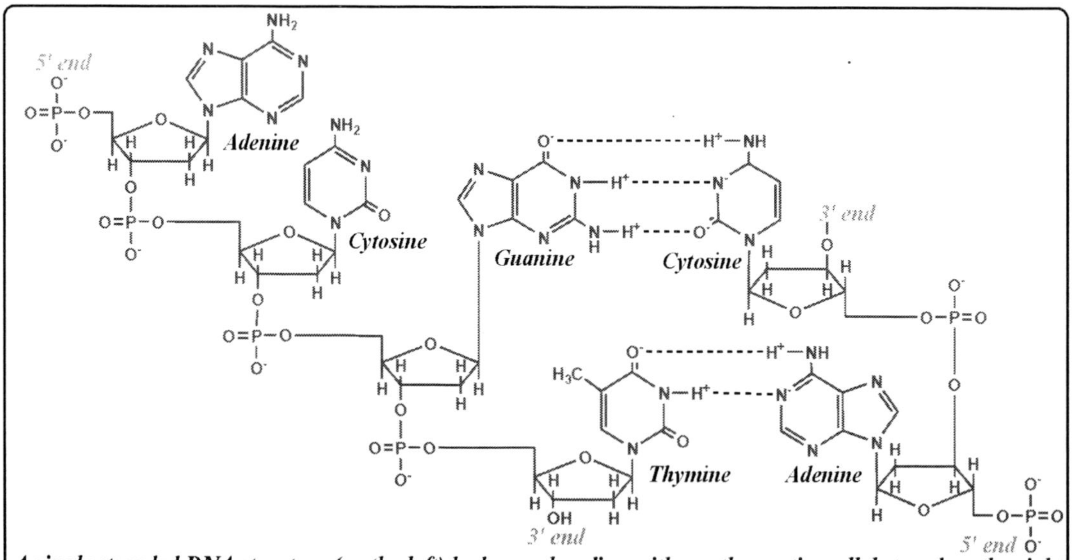

***A single stranded DNA structure (on the left) hydrogen bonding with another antiparallel strand on the right***

Figure 6. The primary structure of the DNA that is partially double stranded to illustrate the nature of the base pairing and hydrogen bond formation.

## The DNA Helix

Helical model of DNA secondary structure was envisioned by James Watson and Francis Crick in 1953. The basic features of the DNA helix model of Watson and Crick are (see Figure 7):

- The DNA consists of two polynucleotide strands wound about a common axis with a right handed twist to form a double helix with ~20 °A diameter (2 nm).
- The two strands run in opposite directions to each other, i.e., they are 'Antiparellel'.
- This conformation leads to the formation of two grooves because the space filling of the adenine-thymine dimer is not the same as that of guanine-cytosine dimer:
  - The major groove is where majority of the regulatory proteins interact with DNA.
  - The minor groove is where histones and very few regulatory proteins interact with DNA.
- In the DNA structure, the hydrophilic deoxyribose and phosphoric acid projects outward forming the backbone of the molecule, whereas, the hydrophobic nitrogenous bases face inward to the core of the molecule to be able to face those from the other strand.
- The plane of the nitrogen bases is perpendicular to the helix axis and the bases are partially overlapping with each other, i.e., base stacking.
- The hydrogen bonds formed between the bases of the opposite strand are 'complementary', i.e., Guanine base pairs with Cytosine by three hydrogen bonds, whereas, Adenine base pairs with Thymine by two hydrogen bonds. These pairs are known as Watson-Crick base pairs.
- The specific association of the base pairs makes the two strands absolutely complementary to each other, i.e., the genetic information is encoded in the sequence of bases on either strand.

*DNA molecules exists in various forms depending upon the base composition and physical conditions; with at least six types (A - E and Z), with A, B and Z forms the most common.*

- *B-DNA* is the naturally occurring right-handed helix with 10 nucleotide pairs per turn, 34 °A turn height (the pitch) and 20 °A double helix width. The B form is the most common form at physiological conditions (low salt and high degree of hydration). Watson and Crick DNA helix corresponds to the B-DNA (see Figure 8).
- *A-DNA* is the dehydrated crystalline form of right handed helical DNA, formed under conditions of lower hydration and higher salt content. This structure has 11 nucleotide pairs per turn, thicker helix than B-form with a diameter of 26 °A and a shorter turn height (28 °A). This form may not exist under physiological conditions. RNA-DNA and RNA-RNA helices exist in similar forms.
- *Z-DNA* is left handed and zigzag-like helix that is thinner (diameter 18 °A) than B-form leading to disappearance of the major groove and deepening of the minor one with 12 nucleotide pairs per turn. The pitch is 45 °A. It is formed under high cation concentration and in areas rich in G and C.

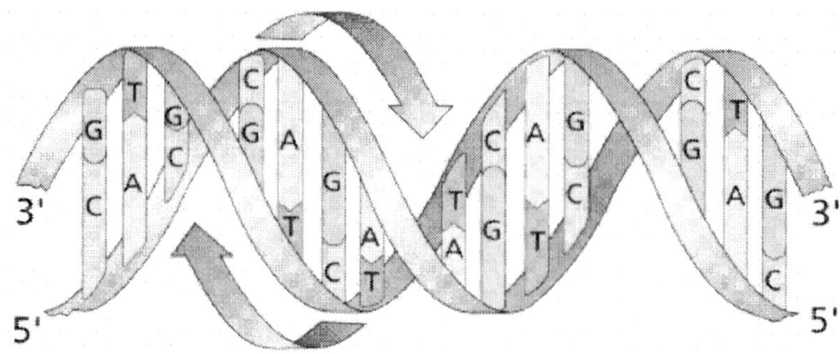


Figure 7. The secondary right-handed helical double stranded form of DNA that creates minor and major grooves.

*DNA may be either single or double stranded and each could be circular or opened/linear:*

- *Double stranded linear chromosomal DNA* is present in nucleus of eukaryotic cells separated from the cytoplasm. It is also present in DNA viruses.
- *Double stranded circular DNA* is present in mitochondria and bacterial chromosomal material and the extrachromosomal plasmid DNA in bacteria and yeast un-separated from cytoplasm. It is also present in plants and some DNA viruses.
- *Single stranded circular DNA* is present in some bacteriophages.
- *Triple-stranded DNA helix* could form *in vitro* and *in vivo*, e.g., during DNA repair and recombination.

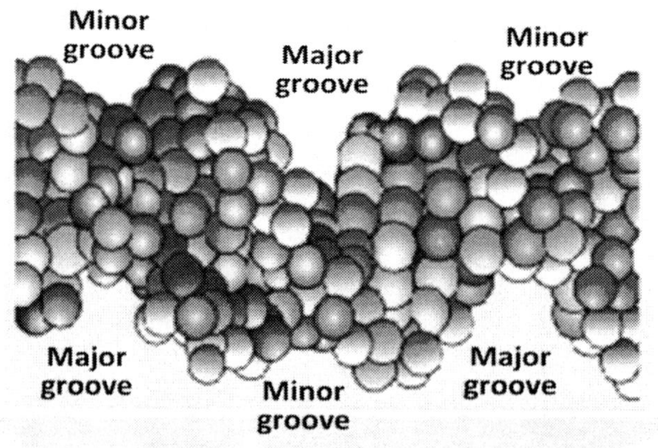


Figure 8. The space-filling B-form of the DNA. External sugar-phosphate backbone (red-yellow) and the internal nitrogenous bases (blue-white) allow formation of the minor and major grooves.

## Chromatin Organization

The genetic material of the eukaryotic cells is enclosed in a membrane bound structure known as nucleus. Nucleus is bound by the bi-layered nuclear membrane with its outer layer in continuation with the endoplasmic reticulum and inner layer supported by the fibrous lamina. The membrane contains pores for exchange of molecules with the cytoplasm. The nuclear fluid phase (karyoplasm) contains proteins, ions, metabolites, lipids and small RNA molecules. During interphase, female somatic cells contain a distinctive mass of darkly stained chromatin close to the nuclear membrane that is the inactive X chromosome. It is compact due to its late replication and inactivation into heterochromatin and is called X chromatin or the Barr body, see Figure 9.

Nucleolus is an area in the nucleus (a suburb) that is not membrane bound and gives dark staining due to association of its boundary with heterochromatin. It is containing the satellite sequences of chromosomes 13, 14, 15, 21 and 22 that contain rRNA encoding genes. Nucleolus is the place of rRNA transcription and the initial assembly of the ribosomal particles. Cell may contain one or more nucleoli, particularly the cells that are active in protein synthesis such as pancreatic acinar cells.

The whole DNA/RNA/Protein (organizing, histone basic proteins and non-histone transcription factors and enzymes) content of the nucleus is called the chromatin and the complete set of the DNA part is called the genome that is $3.3 \times 10^9$ bp in length for human haploid cell.

The cell nucleus varies in several aspects (size, location and shape) depending on cell type and activity, e.g., small, medium or large sized, central, acrocentric, peripheral or basal position, and rounded, rod-like, kidney-shaped, oval flat, or lobulated shape. A cell may be mono-nucleated (most body cells), bi-nucleated (e.g., hepatocytes) or multi-nucleated (e.g., skeletal muscle cells).

The DNA has to be condensed to occupy such small nuclear space. This condensation needs to be in a well-organized manner so as to be easily accessed for transcription, repair and replication. Structure of the genome in such organized form is called chromatin. This organization requires interaction with several proteins and RNAs. Thus, chromatin is formed of double stranded chromosomal DNA, histones (basic protein), non-histone proteins (other basic proteins such as protamines and replication and transcription executors and regulators) and a small amount of RNA. In the interphase, each chromosome is a physically separate one double stranded helical molecule of DNA that ranges in length from about 50 million to 250 million bps.

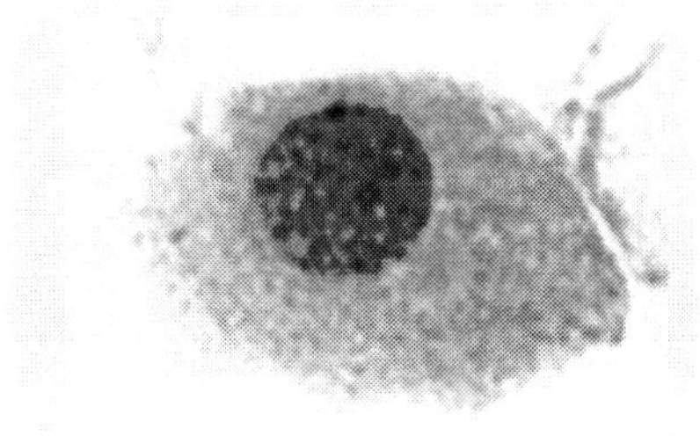

Figure 9. A single female showing the Barr body (the inactive X chromosome) as a dark black spot at the periphery of its nucleus - at 12 O'clock position.

Histones are basic non-heat coagulable proteins that are very rich in lysine and arginine (20-30%). Thus, they are positively charged at the physiological pH to bind the negatively charged DNA molecules (due to -OH of the phosphate groups) by ionic bonds. There are five types of histones: H1, H2A, H2B, H-3 and H4 (ranging from 22-11 kDa MW, respectively).

*Worth noting: The genome is our genetic autobiography*

The genome could be considered as the genetic autobiography book; chromosomes are the chapters; genes are sentences of every chapter; codons are the composite words of every sentence and nucleotides are the letters of these words. The whole length of a haploid human nuclear DNA is about 1 meter (or 3 feet) while the whole cell is 20 μm and nucleus is 5 -15 μm in diameter. Average molecular weight (MW) of sodium salt of a DNA base pair (bp) = 650 Daltons (Da). 1 Da equals the mass of a single hydrogen atom = $1.67 \times 10^{-24}$ grams. MW of a double-stranded DNA molecule = number of bps x 650 Daltons. Total weight of the human haploid genome = $3.3 \times 10^9$ bp x 650 = $2.15 \times 10^{12}$ Da. Since, 1 Da = $1.67 \times 10^{-24}$ grams, so the human genome weighs $3.59 \times 10^{-12}$ grams (i.e., 3.59 picogram). If the entire DNA in a single human cell was placed end to end it would be six feet long. If the entire DNA in all of the cells in a human body was placed end to end it would reach the sun and back 600 times, i.e., 100 trillion cells x 6 feet divided by 93 million miles = 1200 times.

## Nucleosome

The packaging unit of the chromatin is called nucleosome. Nucleosome is formed of a histone core of eight molecules of the core histone barrels-shaped highly package globular octamers; two each of; H2A, H2B, H3 and H4, and, 1.75 super-helical turns (146 - 147 nucleotides length) of DNA wrapped around. The N-terminus of each octamers subunit is protruding free to be accessible for covalent modifications. A linker stretch of DNA (~30 nucleotide in length; 15 - 55 bp depending on the species) connects between the nucleosome cores and is covered by one molecule of histone H1 so as to close two turns of DNA; See Figure 10.

More than the inert structural DNA packaging role, histones in the nucleosome have several dynamic regulatory roles induced by the reversible covalent modifications of the N-terminal tails of the four core proteins (mainly on their lysine and histidine amino acids) by, e.g., acetylation, methylation, ADP-ribosylation, ubiquitinylation and phosphorylation. Their regulatory roles include:

a. Control of rate of gene expression (induction/repression and silencing).
b. Control of chromosomal condensation (hetero- vs. euchromatin) and assembly during replication.
c. DNA repair.

*The nucleosome structure and position are not stable but they are modulable by remodeling, cis-sliding or trans-positioning so as to increase/decrease access to transcription control sequences.*

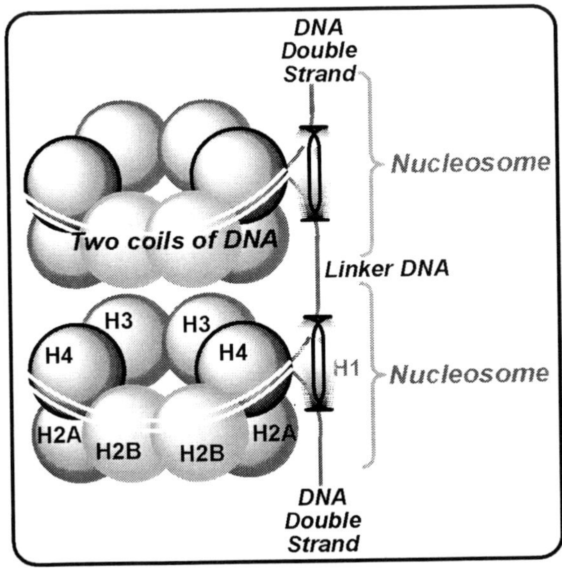


Figure 10. Diagrammatic structure illustrating the main features of the nucleosome. H = histone.

The organization of the 2 nm DNA double strand in nucleosomes makes it to look like beads-in-a-string that is called the 10 nm fibers. This fiber is then coiled again around a linear hollow axis as a helix with 6 - 7 nucleosomes per turn to form the solenoid 30 nm fiber (like the telephone cord). The latter is organized into loops or domains through binding into the nuclear scaffold or matrix proteins in the extended interphase chromosome leading to the 300 nm thickness fiber. Each loop contains 20 -100 kb nucleotides in the euchromatin regions. The looping is brought about through attachment to the nuclear matrix proteins via a specific AT-rich DNA segments called matrix-associated regions or scaffold attachment regions. These regions contain also genetic insulators, and so, delimit a separate functional/structural genetic island (e.g., the β-globin gene cluster). This level of packaging is the maximum condensation for the interphase chromosomes;see Figure 11.

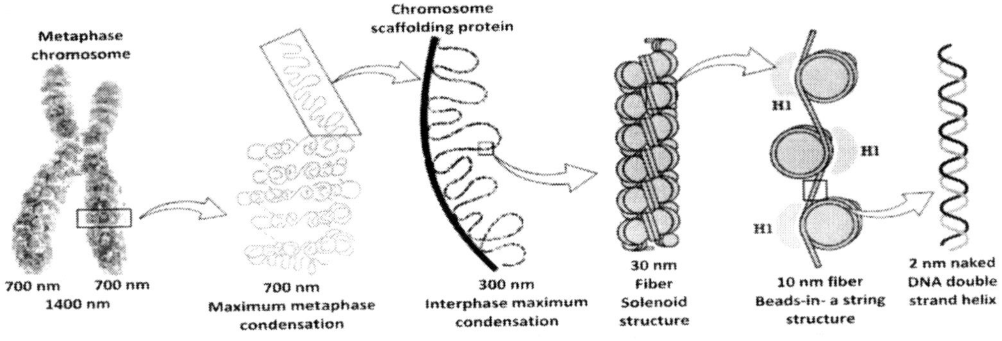


Figure 11. The condensation of the 2 nm thickness DNA double helix - 50,000x times, into the maximum condensation of 1400 nm thickness metaphase chromosome.

With the help of other organizing proteins, the 300 nm fiber form of interphase chromosomes is further condensed into 700 nm thickness sister-chromatid to appear as discrete metaphase mitotic chromosomes. One metaphase chromosome with two sister chromatids will have ~1400 nm thick and is nearly totally transcriptionally inactive heterochromatin. Therefore, when the cell is about to enter mitosis, the linear chromosomal DNA is 50,000x condensed into the short metaphase chromosome.

Chromatin that is transcriptionally active is called *euchromatin* which is accessible for regulatory proteins, has an open conformation due to relaxed state of the nucleosome arrangement, rich with genes, hypomethylated at CpGs, with hyperacetylated at its histones, stains light and is DNase I hypersensitive. But densely packed dark stained, inaccessible chromatin, that is hypermethylated at CpGs, hypoacetylated at its histones, rich with structural repetitive sequences, is not DNase I hypersensitive due to inaccessibility of its DNA, is inactive and is called *heterochromatin*. *Constitutive heterochromatin* is always inactive and found at centromeres and telomeres. When there is heterochromatin DNA that becomes active during specific developmental phase and is inactive elsewhere, it is called *facultative heterochromatin*, e.g., one of the female X chromosomes is almost completely inactive (and is visible condensed as a dark staining "Barr body" pushed against the nuclear membrane) in all female body cells - except for a short period of activity during early embryogenesis.

## **Types of DNA Sequences** (Figure 12)

1. *Unique non-repetitive single-copy gene exon sequences*. They represent 1.1% of the human genome (48 Mb out of the 3200 Mb total haploid genome). The entire content of exons of all genes of an organism is called exome of that organism and the research field investigating it is called exomics. It is the fraction that encodes proteins – compare to >90% of the bacterial genomic DNA that encodes proteins.
2. *Gene-related regulatory, intervening intron sequences and 5'- and 3'-untranslated regions (UTR)*. They represent 42% of the human haploid genome (1152 Mb).
3. *Intergenic repetitive sequences*. They constitute more than 50% of the human haploid genome (2000 Mb) and are repeated at 2 - $10^7$ copies per cell. While the repeated sequences themselves are usually the same from person to person, the number of times they are repeated tends to vary. Some of these stretches of repeats are known as *minisatellites* that have two types; *variable number of tandem repeats* (10-100 bp length and repeated 20-100 time) and *short tandem repeat* (2-9 bp that is repeated more frequently) segregate independently and are heritable in a Mendelian fashion, therefore, can be used for forensic individual *DNA typing or fingerprinting*.

   *Within each chromosome, repetitive sequences* are generally classified into:

   - *Highly repetitive sequences* with 5 - 500 bp length and repeated at 1 - 10 million copies per haploid genome. They cluster at *centromeres* and *telomeres* where they have a structural non-coding role.
   - *Moderately repetitive sequences* with a copy number less than one million per haploid genome. They are *interspersed with the unique sequences*. They could be *long interspersed element sequences* (LINEs) with 6-7 kb length

and 20,000 - 50,000 copies per genome that are species-specific families; or, *short interspersed element sequences* (SINEs)with 70 - 300 bp length and more than 100,000 copies per genome. Example is the species-specific *Alu family* that is transcribed into hnRNA, and, 4.5S and 7S RNA molecules. Alu family is an example of transposons; i.e., DNA sequence that can change position by jumping out and reintegrating into DNA sequence at various sites. This could lead into diseases such *neurofibromatosis*. Homologous recombination between two Alu repeats causes a large deletion in the LDL gene with loss of its expression that leads to *familial hypercholesterolemia*.

- *Microsatellite repeats* that are clustered and interspersed repeats with 2-6 bp length repeated up to 50 times. They are located at 50 - 100,000 locations in the genome, e.g., the AC repeat. Their number may vary in the two homologous chromosomes and is a heritable trait. Their analysis is helpful in mapping genes and detecting linkage to specific disease. Increase of their repeat number than normal, i.e., microsatellite expansion, instability or polymorphism, causes diseases, e.g., unstable p$(CGG)_n$ is associated with *the fragile X syndrome*; $(CAG)_n$ is associated with *Huntington's Chorea*; $(CTG)_n$ is associated with *myotonic dystrophy*; and $(CAG)_n$ is associated with spinobulbar muscular atrophy and Kennedy's disease.

*Worth Noting: DNA Fingerprints*

The full genetic profiles of any two individuals (other than identical twins) reveal many differences on which DNA typing relies. These differences are mainly in the repetitive *minisatellites/microsatellites*. Short tandem repeats are more useful because they are dispersed more evenly throughout the genome than the longer variable number of tandem repeats. While the repeated unit sequence is usually the same from person to person, the number of times they are repeated at specific chromosome locus tends to be individual-specific. The number of repeats at a particular locus can be calculated by dividing the entire molecular weight of a given marker by the molecular weight of the repeated unit sequence. Using a number of different loci markers, e.g., 13, makes the profile $1/10^{12}$ individual-specific for forensic, identity, and paternity uses. see the applied molecular biology section.

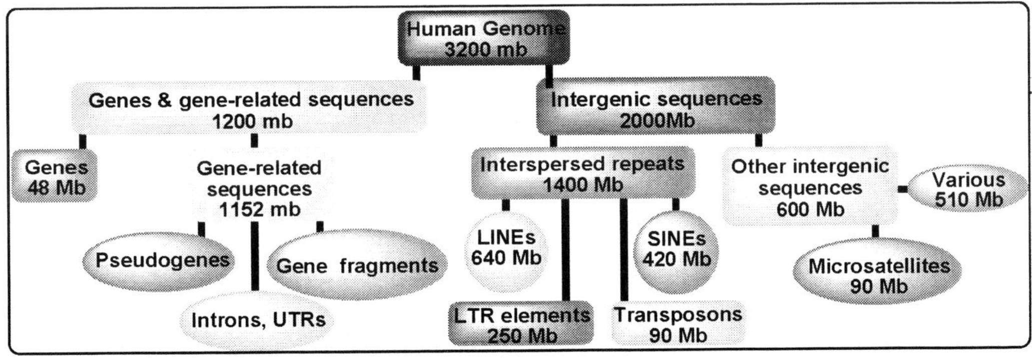


Figure 12. Types of DNA sequences. UTRs = un-translated regions; LINEs = long interspersed element sequences; SINEs = short interspersed element sequences; and LTR = long terminal repeats.

Normal human somatic cells have 46 chromosomes: 22 pairs (or *homologs*) of autosomes (chromosomes 1-22) and two sex chromosomes. This is called the *diploid* number. Females carry two X chromosomes (46, XX) while males have an X and a Y (46, XY). Germ gamete cells (ova and sperms) have 23 chromosomes: one copy of each autosome plus a single sex chromosome. This is referred to as the *haploid* number. Thus, one chromosome from each autosomal pair plus one sex chromosome is inherited from each parent. Mothers can contribute only an X chromosome to their children while fathers can contribute either an X or a Y, and so, baby sex determination is the male duty - although it requires a permissive uterine environment. Chromosome 1 contains 263 Mb of DNA while chromosome 21 has 50 Mb. Some chromosomes ARE gene-rich (e.g., 19, 22) while others are gene-poor (e.g., 4, 18).

All body cells contain same genetic information (except mature plasma B-cells) and difference between cells is due to differential expression (usage) of genes. There are ~25,000 polypeptide-coding genes in each human cell. The remaining non-coding DNA sequences are as important as the non-repetitive gene-coding sequences and carry out structural, regulatory and identity functions. Liver and kidney cells express 10,000 - 15,000 genes out of the total ~25,000 genes as an example of the tissue-specific expression of genes.

## The Human Metaphase Chromosome

Under the microscope *metaphase chromosomes* possess a two-fold symmetry with the identical duplication of sister chromatids. Each chromatid contains one DNA molecule in a super-coiled from with organizing proteins. The chromatid is structurally differentiated into a short arm and long arm separated by a primary constriction called the *centromere*. The short arm is designated as *p* (*petit*) and the long arm as *q* (*queue, grand*). Centromere is composed of AT-rich DNA sequence ranging in size from $10^2 - 10^6$ bp of constitutive heterochromatin that is made complex with several specific proteins to from *kinetochore* at which mitotic spindle anchor during segregation of chromosomes for mitosis. For replication each chromosome needs to have its centromere, telomeres and origin(s) of replication sequences. Metaphase chromosomes are numbered in order of size from largest 1 to smallest 22. However, 21 is actually smaller than 22.

*DNA-specific stains*, e.g., Giemmsa's generate specific banding due to specific DNA/protein interaction. The banding is not random but is species-specific and is very reproducible among individuals of the same species. Each stained metaphase chromosome arm is divided into chromosome-specific regions labeled p1, p2, p3 etc., and q1, q2, q3, etc., counting outwards from the centromere. Regions are delimited by specific landmarks, which are consistent and distinct morphological chromosome features, such as the ends of the chromosome arms, the centromere and certain prominent bands and secondary constrictions. Regions are divided into bands labeled p11 (one-one, not eleven!, i.e., band one-region one), p12, p13, etc., sub-bands labeled p11.1 (i.e., subband one-band one-region one), p11.2, etc., and sub-sub-bands, e.g., p11.21 (i.e., sub-sub-band one-subband two-band one-region one), p11.22, etc., in each case counting outwards from the centromere. Thus, 9q34.12 means the $2^{nd}$ sub-sub-band of the $1^{st}$ subband of the $4^{th}$ band of the $3^{rd}$ region of the long arm of chromosome 9. The sub-banding and the sub-sub-banding will be more easily resolved when the chromosomes for staining are collected from dividing cells within the G2 phase of the cycle before complete compactation to the metaphase final size. The centromere is designated '*cen*' and the telomere '*ter*' when describing any abnormalities related to them. *Proximal* Xq means the segment of the long arm of the X that is closest to the centromere, while *distal* 2p means the portion of the short arm of chromosome 2 that is most distant from the centromere, and therefore closest to the telomere. The diagram that shows these chromosomal specific banding patterns and details is called *ideograms*; a "chromosome map" showing the relationship between the short and long arms, centromere, the stalks and satellites. The band numbering helps describing chromosomal rearrangements.

Human metaphase chromosomes come in three basic shapes and can be categorized according to the length of the short and long arms and also the centromere location. *Metacentric* chromosomes have short and long arms of roughly equal length with the centromere almost in the middle. *Submetacentric* chromosomes have short and long arms of unequal length with the centromere more towards one end. *Acrocentric* chromosomes have a centromere very near to one end and have very small short arms. Arrays of repeated ribosomal DNA genes on the short arms of the acrocentric chromosomes 13, 14, 15, 21 and 22 often appear as secondary constriction thin stalks (*st*) carrying knobs of chromatin; satellites (*sa*) connected to the centromere. The following are ideograms of Giemmsa's-banded chromosomes 1, 9, and 14 as typical examples of metacentric, submetacentric, and acrocentric chromosomes, respectively. see Table 2.

**Table 2. Classification of chromosomes according to the position of the centromere. Examples showing regions, bands, and sub-bands for chromosomes 1, 9 and 14**

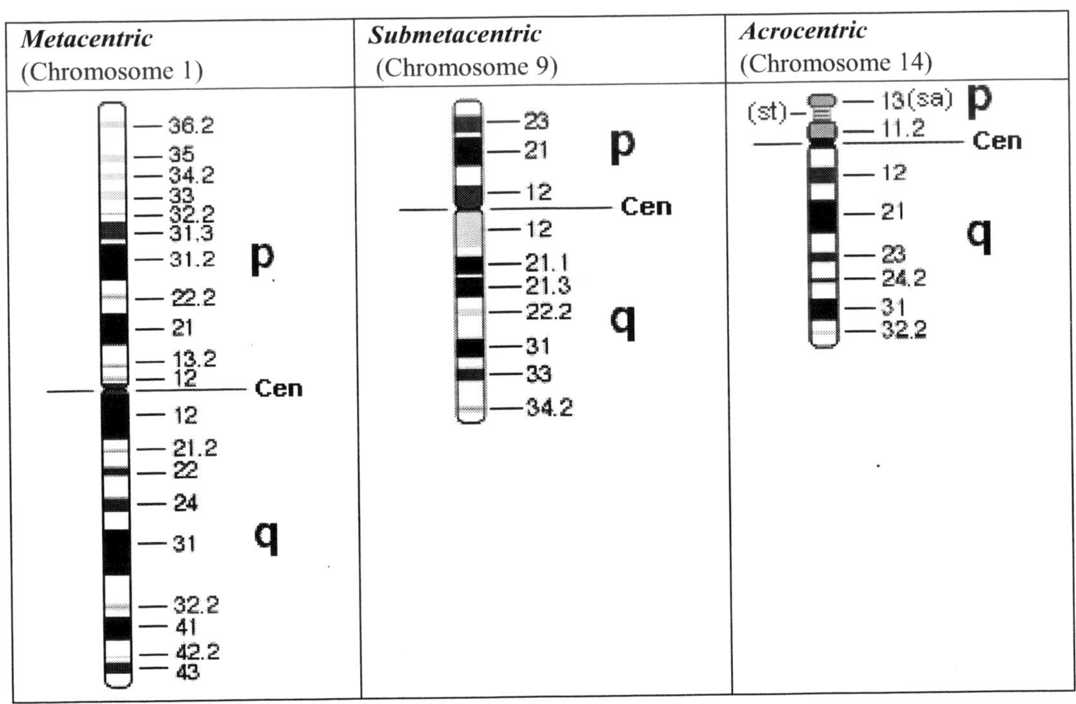

# DNA Replication (DNA Synthesis

Replication is the process of making a 'replica' of the DNA. The duplication of DNA double stranded helix is a must before the mitosis that involves segregating a copy of the duplex into the new daughter cells. Since DNA is a double helix, each old strand will be copied into a complementary new strand so that during segregation each new duplex would have one old and one new DNA strand. Therefore, replication is semi-conservative in nature, which means that each daughter cell will receive a DNA strand from the mother cell and a complementary newly-synthesized strand.

*Worth Noting: Watson and Crick Assumption*

The exact words of James D. Watson and Francis H. C. Crick in 1953, envisioning the semi-conservative nature of replication, were as follows, *"Now our model for DNA is, in effect a pair of templates, each of which is complementary to the other. We imagine that prior to duplication the hydrogen bonds are broken, and the two chains unwind and separate. Each chain then acts as a template for the formation onto itself of a new companion chain, so that eventually we shall have two pairs of chains, where we only had one before. Moreover, the sequence of the pairs of bases will have been duplicated exactly."*

The semi-conservative nature of the DNA replication proposed by Watson and Crick was confirmed by an elegant experiment by Meselson and Stahl (1958) using two isotopes of nitrogen. In their experiment they first allowed E. coli to grow for a number of generations in a medium containing $^{15}N$, so that the entire DNA was labeled with the heavy nitrogen isotope. Then these bacteria were transferred to another medium containing the light $^{14}N$ isotope and allowed to divide. The bacterial samples were harvested at different time periods and DNA extracted from the cells was subjected to density-gradient equilibrium sedimentation. The results were correlated with the doubling time of the bacterial cells. First generation cells showed the existence of DNA duplexes that had a 'mixed' density, i.e., they were hybrids of $^{15}N$ and $^{14}N$ DNA strands. The second generation showed the presence of pure $^{14}N$ labeled and the hybrids in equal quantity. In the subsequent generations, the proportion of the hybrids continued to decrease and that of pure $^{14}N$ labeled DNA increased. These results could only be explained on the basis of a 'Semi conservative' type of replication.

## Enzymes Participating in DNA Replication

The basic process of replication has been well studied in prokaryotic systems but the eukaryotic systems have also been found to follow the same principles. DNA always reads from 3'-end to 5'-end for replication and transcription. The process involves a number of enzymes and proteins.

*The prokaryotic replication system:* The following list of enzymes is not exclusive and about 10 more proteins are required. Generally, the need of so many proteins in the process is mainly due to the synthesis and subsequent removal of a primer - small stretch of 'preformed' DNA or RNA with a free 3'-OH - upon which DNA polymerases synthesize DNA. This is because DNA polymerases cannot start DNA replication *de novo* on their own.

- *DNA polymerase I*, first replication enzyme to be isolated by Arthur Kornbreg and hence known as *Kornberg's enzyme*; is responsible for DNA repair and gap filling by nick translation of the RNA primers into DNA sequence with help of RNaseH. The enzyme has 5'⇒3' DNA polymerase activity for polymerization and both 3'⇒5' and 5'⇒3' exonuclease activities for 'proof reading'.
- *DNA polymerase II* also has 3'⇒5' proofreading and DNA repair function.
- *DNA polymerase III* is the major enzyme responsible for DNA replication and has 3'⇒5' proofreading. It encircles the DNA strand with the help of the sliding β2 clamp complex and the γ loader clamp complex.
- The *clamp loading γ-complex* that loads the *β-sliding clamp*.
- *DNA helicase* unwinds the double helix into two single-stranded DNA. It breaks down hydrogen bonds between the bases at the expense of ATP. Two ATP molecules are consumed for each base pair breakage. Once separated, single stranded DNA-binding proteins stabilize the single stranded DNA formed.
- *The single-stranded DNA binding proteins (SSBP)* prevent reannealing of DNA helix by binding the single stranded DNA created by helicase during replication, recombination and DNA repair. They do not interfering with DNA polymerase function. The eukaryotic counterpart is the replication protein A (RPA). These binding proteins also protect DNA from forming the replication hindering stem-loops or being degraded by nucleases.
- *Primase* (DnaG) is a DNA-dependent RNA polymerase that synthesizes short RNA primers (10-200 bases) in the 5'⇒3' direction with the help of DNA binding protein and helicase (primosome). RNA primer is complementary and anti-parallel to the DNA. The primer has a free 3'-OH to accept deoxy-ribonucleotides polymerized by DNA polymerase III.
- *DNA gyrase* removes positive supercoils by introducing a cut in the DNA duplex utilizing energy from ATP and introduces negative supercoils before sealing the duplex again.

**Table 3. Comparison of prokaryotic and eukaryotic DNA polymerases**

| Function | Prokaryotes | Eukaryotes |
|---|---|---|
| DNA nick translation and repair | Polymerase I (+ proofreading) | Polymerase α (high fidelity) |
| DNA replication leading strand | Polymerase III (+ proofreading) | Polymerase δ (high fidelity) |
| DNA repair | Polymerase II (+ proofreading) | Polymerase ε (high fidelity) |
| DNA repair | — | Polymerase β (low fidelity) |
| Mitochondrial DNA replication | — | Polymerase γ (+ proofreading) |

- *DNA ligase* joins the free 3'-OH on deoxyribose-C3 of one end nucleotide to 5'-OH of phosphate on deoxyribose-C5 of the other end nucleotide of DNA. It utilizes NAD and ATP as the energy source in prokaryotes and eukaryotes, respectively.

*The Eukaryotic replication system:* The eukaryotic DNA is very large, separated into chromosomes and is associated with nuclear proteins as nucleosomes. Therefore, eukaryotic replication is more complex than the prokaryotic one and involves much larger number of proteins and enzymes. It requires displacement of histones from DNA.

- *Mammalian DNA polymerases:* The five major DNA polymerases take part in replication are; DNA polymerase-α, -β, -γ, -δ, and -ε. DNA polymerase α carries out function of prokaryotic DNA polymerase I but along with its RNA primase activity. It has a DNA polymerase activity, too, that enables it to add DNA nucleotides on the RNA primer. DNA polymerase β and ε are concerned with DNA repair with low processivity. DNA polymerase ε has proofreading 3'⇒5'-exonuclease activity and may replace DNA polymerase δ in replicating the lagging strand. DNA polymerase δ is formed of two subunits. The enzyme interacts with the proliferating cell nuclear antigen (PCNA) that increases its processivity. It carries out function of DNA polymerase III as the major replicating enzyme and has proofreading 3'⇒5'-exonuclease activity. DNA polymerase γ is the mitochondrial DNA polymerase with high fidelity and proofreading 3'⇒5'-exonuclease activity. see **Table 3** for major differences between prokaryotic and eukaryotic DNA polymerases. There are other types of eukaryotic DNA polymerases that are concerned with DNA repair.
- *Topoisomerase I:* During separation of the double stranded DNA, positive supercoiling of the helix occurs in front of the fork due to accumulation of the torsion force which makes further progress of replication fork difficult, while, negative supercoils occur behind it. Instead of rotating the whole chromosome ahead of replication fork into the opposite direction that requiring higher amount of energy, topoisomerase I nicks (cut) a single DNA strand of the helix. Each end of this strand is free to rotate in the opposite direction of the supercoils to relax the helix by passing the intact strand through the nick. Then, the enzyme seals the cut again with no expenditure of energy.
- *Topoisomerase II* (equivalent of the gyrase in prokaryotes) cut the DNA duplex to relax the supercoils on the expense of ATP for energy. The free ends rotate till relaxed and then the enzyme seals the duplex without introducing negative supercoils (compare with gyrase).
- *The sliding and loading complex clamps* of the prokaryotes are replaced by the replication factor C (RFC) and the PCNA in eukaryotes.

*Worth Noting: Supercoiling*

*Positive supercoiling:* Unwinding of the helix of the circular and linear DNA with fixed ends creates stress that twists the right-handed DNA helix around its axis in the clockwise direction of its turns and will contain more than 10 base pair per turn. Oppositely, *negative supercoiling* occurs in such molecules due to twisting the right-handed DNA helix around its axis in the anti-clockwise direction of its turns leading to straining of the helix and disruption of hydrogen bonds at small regions. Negative supercoiling is preferred in the biological system because it facilitates the separation of the DNA helix into single strands during DNA replication, repair, and transcription.

## Sites of DNA Replication

DNA duplex in prokaryotes has a limited length, so replication starts at a single AT-rich site at which the two strands separate that is called origin of replication (OriC or replicon). This site has specific nucleotide sequences that are recognized by enzymes and proteins responsible for initiation of DNA replication (collectively called origin recognition complex; ORC).

Since the eukaryotes DNA duplex is too long, replication has to start at several points along one molecule to be finished within the limited S phase time (~9 hrs) and to protect the naked DNA from mutations. For that replication starts at several hundreds of replicons a long each chromosome. The eukaryotic origin of replication (replicon) is repeated along the chromosome every 1-300 kbp and each replicon sequence span 100 - 50,000 bp. Origin recognition complex (ORC) is a multi-subunit protein that binds replicon and is a docking site for CDK4/6 at G1-S transition. The replication licensing factor or MCM (*M*ini-*C*hromosome *M*aintenance) protein binds the DNA sequences flanking the replicon. The cell cycle regulatory kinase, cyclin D-dependent kinase 4/6 (D/CDK4/6) phosphorylates and releases the S phase CDK2 to bind cyclin A. The active cyclin A-CDK2 phosphorylates and releases MCM so as the replicon becomes accessible by the ORC for initiation of replication; see control of the cell cycle.

The double stranded DNA separation at each replicon creates replication bubbles. Since each end of the bubble is a V-shaped structure, it is called a replication fork that progresses outwards from the replicon. see Figure 13. Replication proceeds in both directions from these replicons simultaneously to guarantee rapid replication of the long DNA duplex.

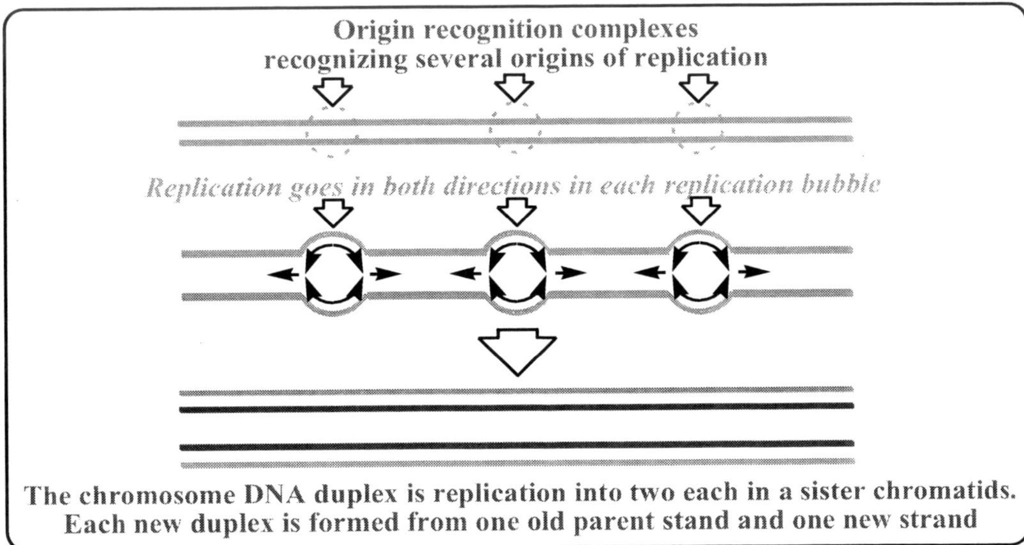


Figure 13. The multiple origins and replication bubbles formation hastens the rate of replication of the very long eukaryotic DNA molecules.

## Steps of Replication

Replication needs the building units of deoxynucleotides, i.e., dATP, dGTP, dTTP, and dCTP in well-balance amounts. Energy-wise, it is a highly expensive process. When all the resources of the DNA replication are available replication proceeds as follows:

- The OriC site is identified by a group of proteins (e.g., dnaA and dnaC in prokaryotes) that separate DNA helix into localized single-stranded DNA ATP-dependently. ATP-dependent helicase (dnaB in prokaryotes) unwinds the DNA duplex. This process results in positive supercoiling in DNA ahead due to accumulation of the torsion force of the helix.
- Binding of single-stranded DNA binding protein (SSBP in prokaryotes) stabilizes DNA in a single-stranded form. The positive supercoiling is relieved by DNA gyrase and topoisomerases with introduction of negative supercoils ahead of replication fork.
- Primase (or polymerase α) synthesizes 10 - 200 bp long RNA primer by polymerizing ribonucleotides in the 5'⇒3' direction from the OriC site according to the base pairing rule: A-U and G-C using the 3'⇒5' DNA strand as the template. The mobile helicase/primase complex is called primosome. The primase activity of DNA polymerase α synthesizes 10 nucleotides RNA primer upon which it adds 10-20 DNA nucleotides by its DNA polymerase component.

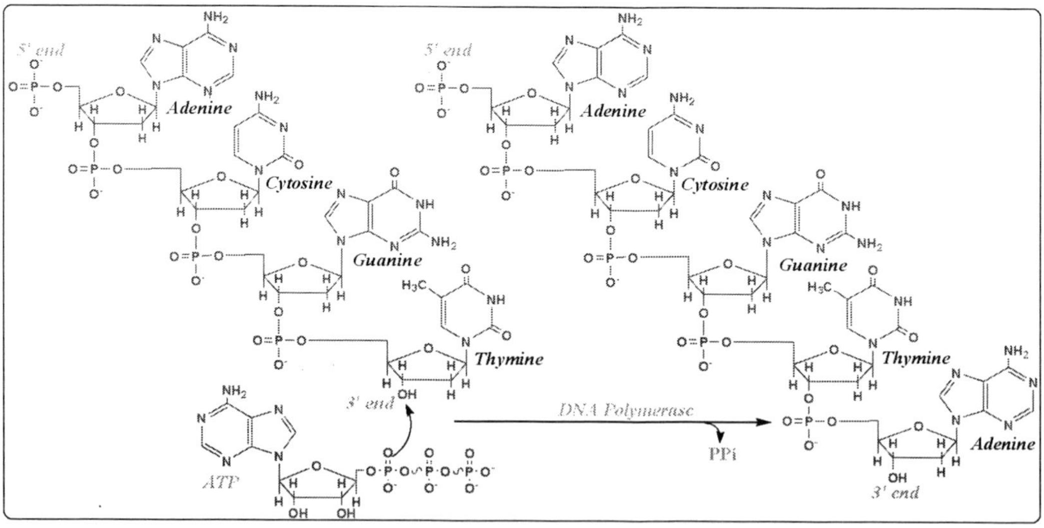

Figure 14. The 5'⇒3' direction of polymerizing DNA nucleotides with the release of the two peripheral phosphates each time (PPi).

- RFC assembles PCNA at the end of the primer to replace DNA polymerase α for DNA polymerase δ that binds the RFC-PCNA complex at the growing 3'-ends to complete the highly processive DNA synthesis. DNA polymerase III and δ add free deoxy-ribonucleotides to the free 3'-OH end of the RNA primer in the 5'⇒3' direction in a continuous manner (*the leading strand*). This is done according to the base-pairing rule: A-T and G-C using the 3'⇒5' DNA strand as a template: see Figure 14.

- The other 5'⇒3' strand, if replicated simultaneously, would require the new DNA strand to be polymerized in 3'⇒5' direction. But no such polymerase is known to exist. Therefore, replication on this 5'⇒3' strand (*the lagging strand*) lags behind for a while until at least 1000 to 5000 nucleotides in prokaryotes (but only 150 - 250 nucleotides in eukaryotes = size of a nucleosome) are exposed. A long with the fact that histone as nucleosomes coats eukaryotic DNA guard against eukaryotic DNA mutation. The primase (or polymerase $\alpha$) makes a primer at the upper end of the replication fork, i.e., in 5'⇒3' direction as in the leading strand. DNA polymerase III extends the primer with DNA nucleotides to cover the downstream sequence. This process is repeated when enough stretch of DNA is again exposed due to the continuous progress of the leading strand. The primase makes a new primer that is elongated by DNA polymerase III until it reaches the previous RNA primer, and so on. The discontinuous synthesis of DNA on the lagging strand gives rise to a number of small stretches of newly synthesized DNA. Each small stretch of DNA synthesized by DNA polymerase III in between the two primers on the lagging strand is called Okazaki fragment. Lagging strand replication dilemma was resolved by Okazaki. DNA polymerase $\delta$ works the same on the lagging strand.
- When the DNA polymerase $\delta$ approaches the RNA primer of the downstream Okazaki fragment, RNaseH1 removes all but the last RNA nucleotide of the RNA primer and the flap endonuclease 1/RTH1 exonuclease complex removes the last RNA nucleotide. DNA polymerase $\delta$ continues filling in the created gap and DNA ligase joins the Okazaki DNA fragment to the growing strand. The prokaryotic RNA primers are removed by the 5'⇒3' nick translation (exonuclease) activity of DNA polymerase I and are replaced with DNA nucleotides. The 5'-phosphate-end of DNA fragments built by DNA polymerase III is linked to the 3'-OH-end of the DNA fragment built by the DNA polymerase I by ATP-dependent *DNA ligase*.

Thus, one strand is built in one patch towards the replication fork 5'⇒3', i.e., the *leading strand*, while the other strand is built in the opposite direction but also in 5'⇒3', i.e., *lagging strands*, in discreet Okazaki fragments; see Figure 15. The proof-reading (i.e., an enzyme catalyzes a reaction in a first step and then checks that the product is correct in a second step) 3'⇒5'-exonuclease 3'⇒5'-exonuclease properties of DNA polymerases ensure a mis-incorporation frequency of one in every $10^8$-$10^9$ bp synthesized. The mismatch repair mechanisms improve mis-incorporation rate to $1/10^{10} - 10^{11}$. Along with DNA polymerases, eukaryotic proof-reading is provided by other enzymes as a repair process. DNA helix reassembles afterwards into nucleosomes and the two strands are differentially methylated (the old strand is hypermethylated than the new one).

DNA methyltransferases methylate CpG sequences at the 5'-position of the cytosine in the newly synthesized DNA strand in the same pattern like the parental strands, therefore copying the DNA methylation pattern from template to daughter strands. Knockout of the DNA methyltransferase genes is lethal embryonically.

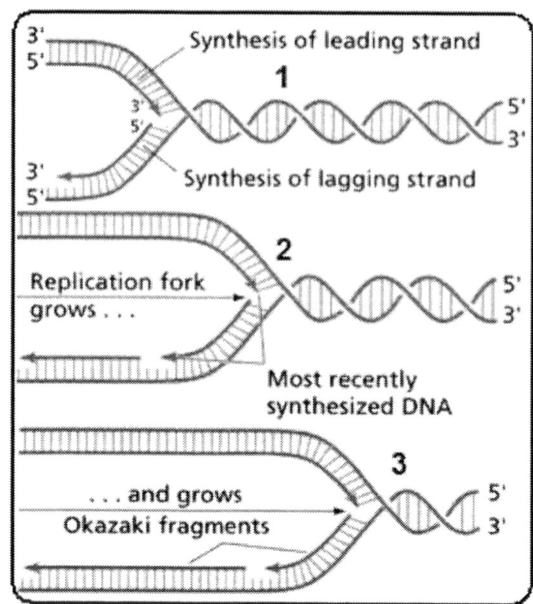


Figure 15. The progress of one replication fork illustrates the difference between the continuously synthesized leading strand vs. the discontinuously synthesized lagging strand.

*Worth Noting: Inhibitors of DNA Synthesis*

Progress of replication can be stopped by the presence of a number of inhibitors. *Actinomycin D* (Dactinomycin) and *daunomycin* intercalates in the major groove of DNA and prevents both replication and transcription. *Topoisomerase inhibitors* such as Etoposide are used as anticancer chemotherapy. *Inhibitors of the prokaryotic DNA gyrase* are used as antibacterial agents such as Quinolones, e.g., Ciprofloxacin. *Nucleotide analogs* with modified sugar and/or bases are competitive inhibitors of DNA polymerases. Chain terminating nucleotides (contain C2 and C3 Dideoxy sugar) and inhibitors of synthesis of nucleotides are used as anticancer, immune suppressive and antiviral chemotherapy. Likewise, synthetic analogs of purines, pyrimidines, nucleosides, and nucleotides with altered heterocyclic ring or different sugar residue are used to inhibit DNA replication, disrupt base pairing and revent cell division for cancer chemotherapy and immune suppression against rejection of a transplanted tissue. Examples include; 5-fluoro- or 5-iodouracil, 3-deoxyuridine, 6-thioguanine and 6-mercaptopurine (they contain an -SH on C6), 5- or 6-azauridine, 5- or 6-azacytidine and 8-azaguanine (aza = N replaces C at that position), cytarabine (contain arabinose rather than ribose), and azathioprine that is converted into 6-mercaptopurine.

## Telomere and Telomerase

*Telomere* is the double stranded non-coding gene-less deoxynucleotides repeats at each end of eukaryotic chromosomes with the associating proteins, *Telomeric Repeat-binding Factors*. It is essential for maintaining the coding sequences at ends of eukaryotic chromosomes from progressive corrosion by exonucleases. Each repeat is a short nucleotide sequence added by the enzyme known as telomerase. *Telomerase* is a ribonucleo-protein reverse transcriptase (RNA-dependent DNA polymerase) complex that is composed of a number of proteins and a non-coding telomeric RNA molecule.

In the human enzyme, the RNA component is 450 nucleotides in length and contains near its 5'-end the sequence 5'-CUAA*CCCUAA*C-3', whose central 6 nucleotides are the reverse template complement of the human telomere tandem repeat sequence $(5'\text{-TTAGGG-}3')_n$ that is added on the 3'-ends of each DNA strand of each chromosome in 5'⇒3' direction *during interface of the human cell cycle*. The repetitive addition of this repeat reaches more than 1,000 times. Then, the DNA polymerase δ uses the telomere sequence as template with and without an RNA primer to make a complementary sequence in 5'⇒3' direction till reaching the 5'-end of the other DNA strand and nick is sealed by DNA ligase. Thus, DNA at the telomere is almost double stranded.

Telomeric DNA loops back on itself to form a circular structure that is stabilized by base pairing between a short stretch of a single-stranded telomere with the double strand GG base pairing to seal and protects the ends of chromosomes. Once synthesizes, the telomeric DNA is replicated in the normal fashion during DNA replication. But the continuous telomere length shortening by exonucleases requires continuous balancing synthesis by telomerase and protection by the telomeric repeat-binding factors; see Figure 16.

Although the initial maximum length of the telomere is a controlling factor for the replicative lifespan of a cell (population doublings), the rate of its loss per cell division is more influential. Thus, although telomere may be longer in short-living mammals such as mice, it shortens more rapidly than in long-lived ones. Prior to complete erosion of the telomere a signal is sent to p53 protein and cyclin-dependent kinase inhibitors (CDKIs, see the cell cycle later) to stop the cell cycle and induce a slow-decaying, non-replicative state known as replicative senescence. After the complete loss of telomeres, typically a cell will invoke apoptosis ("cell suicide") or become replication senescent to prevent dividing and acquiring mutations that could lead to cancer. Defects in the telomere maintaining proteins, i.e., Telomeric Repeat-binding Factors, and/or induction of a high telomerase activity lead to chromosome instability that predisposes to cancer.

Telomere length varies widely between individuals, cells of the same tissue and among chromosomes of the same cell. The critical interaction of telomere with cell-cycle regulatory proteins reflects its impact on the cell "mitotic clock". At conception each human telomere is ~10,000 - 25,000 bp long. Due to the average loss of eight TTAGGG subunits (a total of ~48 bp) per cell division (equivalent to an RNA primer; 40 - 120 bp) from the 5'-end of each new DNA strand each cell division cycle, nine months later, at birth, the average telomere is half as long as it was at conception. Telomerase function and telomeric sequence are particularly important for proliferating cells (zygote, and stem and germ cells) with highest telomerase activity, than non-dividing cells such as nerve and muscle fiber cells. Figure 17 compares the relative length of telomeres in a young replicative stem cell and a non-replicative senescent cell.

Telomerase knockout mice maintain telomere length by a mechanism known as *Alternative Lengthening of Telomeres* that also occur in human cells but at a much less frequency. This mechanism is induced by mutant p53 function and utilizes homologous recombination system of DNA repair. For humans, the length of the remaining telomere is usually an indicator of how many divisions a dividing cell has left. This explains in part one of the serious disadvantage of animal cloning, i.e., premature aging, since the nuclear donor cell will have a higher chronological age compared to natural zygote and, hence, the cloned animal will be an aged animal at birth.

DNA sequences; and the purple is the DAPI (4,6-diamidino-2-phenylindole) DNA stain.

- *Prevention of recombination* with foreign DNA at the ends of the chromosome and prevents formation of dicentric chromosomes due to fusion of the ends of two chromosomes. However, cross-over is relatively common at the telomeric regions than close to centromeres. Moreover, loss of telomere appears to affect the ability to repair DNA double strand breaks.
- *It controls the rate of the regional sub-telomeric gene transcription* through interaction between the telomeric ends of chromosomes and the chromosome/nuclear matrix.
- Telomeric shortening to a critical length leads to loss of sub-telomeric genes, genomic instability; improper DNA repair and stoppage of cell dividing and induction of apoptosis. This is the rationale behind one of the *theories explaining aging*.
- Telomerase is reactivated to a high level in 90% of human malignant cells that allow indefinite cell proliferation and immortalization. Therefore, telomerase inhibitors and its gene disruption are rational *anti-cancer therapeutic approaches*. Transfection of human somatic cells (e.g., fibroblasts, retinal epithelial and vascular endothelial cells) with the telomerase expressing gene increases their maximum population doubling into 70 times with normal, healthy appearance and activity. However, this makes cells more resistant to apoptosis induced by oxidative stress. Oppositely, to critically short telomeres triggers a p53-regulated DNA-damage checkpoint response, cell cycle arrest and apoptosis. *In vitro* inhibition of telomerase activity leads to tumor cell apoptosis.
- Inherited abnormally short telomeric length and/or abnormally enhanced telomeric loss cause a number of hereditary aging and cancer predisposing syndromes that include: Ataxia Telangiectasia; Down's syndrome; Hutchinson-Gilford Progeria Syndrome, and Werner's syndrome.

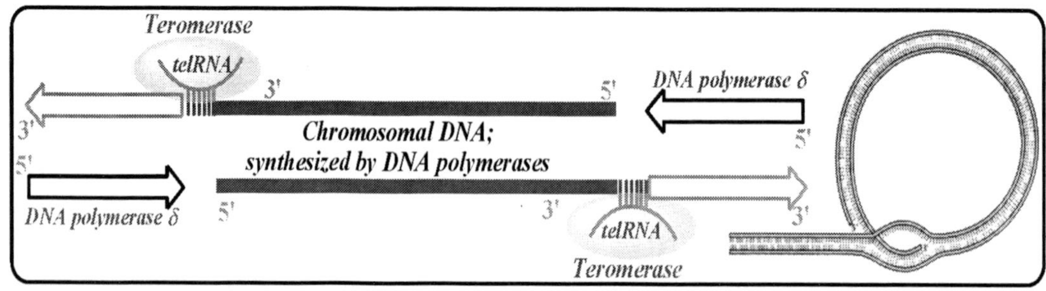


Figure 16. The telomeric DNA is synthesized by the concerted action of telomerase and DNA polymerase (on the right). Telomere loops to seal the ends of the chromosome (on the left).

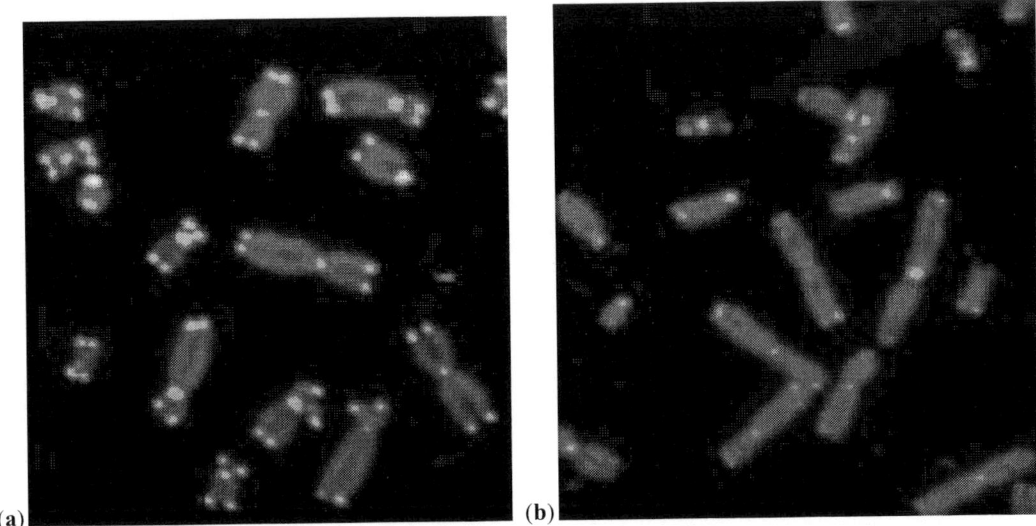


Figure 17. Fluorescence in situ hybridization (FISH) for a stem cell with long telomeres (a) and for a senescent cell with very short telomeres (b). The green-tagged probe hybridizes the telomeric DNA sequences; the pink-tagged probe hybridizes the centromeric Cell Cycle and its Regulation.

A cell is the fundamental unit of life and, hence, a living cell has all the characteristics of life,; i.e., it arises from a living cell, gets differentiated into a range of types of cells with diverse functions, thrives in and responds to the changes in its environment, propagates life by giving birth to a new cell, and comes to its end after a specific lifespan. The progress of a cell from a new life to its giving rise to new cells is known as the cell cycle.

Cell cycle is composed of the highly regulated phases cell goes through to duplicate its DNA and transmit genetic information into two new daughter cells. *Doubling or generation time* is the time utilized by a cell to complete a single cell cycle. The generation time shows considerable variation from less than 24 hours (10-20 hrs) to a few days (up to 100 days) depending on; cell type, developmental state, the growth conditions, tissue type, animal age and species. Each normal cell type has a maximum number of *population doublings in vitro* before cell dividing ceases, i.e., become replication senescent, e.g., embryonic fibroblasts divide about 50 times, whereas, blood progenitor cells and immortal malignant cancer cell are theoretically immortal ever-dividing.

The cell cycle is an inherent character of each cell type that correlates its maximum lifespan and is controlled mainly through the length of their telomere and telomerase activity. However, this 50-doubling limit does not correlate well with the whole organism lifespan. Fibroblasts from a 12-year old chicken has 25 doublings, whereas, those from a 175-years old Galapagos tortoise has 130 doublings.

## **Stages of the Normal Cell Cycle** (Figure 18)

Cell cycle is initiated by the arrival to the quiescent cell of a growth-promoting signal that could be cell-cell, cell-matrix or a soluble growth factor/receptor interaction. Grossly, cell cycle may be divided into two main phases: namely, Mitosis and Interphase. The *mitotic or M phase* is the shorter phase, lasting 1 - 2 hours, during which cell divides into two daughter cells after chromosomal separation. The *interphase* is the period of growth and preparedness between two M phases that lasts for 16 - 24 hours and includes 3 phases; namely, G1, S, and G2. Cell roughly doubles it mass and DNA content (from diploid into tetraploid) during the interphase. Thus, a cell cycle is composed of four phases; G1, S, G2 and M. The non-dividing cells temporarily or permanently lodge at Go. Although the most spectacular morphological changes are observed in the M phase, there are highly active biochemical activities during the other phases.

*G1, Growth or Gap phase 1:* Once cell was born by a preceding mitotic division, it has one of three fates;:1) to grow gradually to an optimum size, 2) to go into a new cell division if it already having such optimum size, or, 3) to withdraw from the cycle into *Go*. The control on cell division is exerted primarily in the G1 phase. G1 phase is variable in length and is responsible for variation in doubling time among different types of cells and average to 10 hr. During G1 phase, cellular DNA content is diploid and chromosomes are in the form of single chromatid and are thin and extended. Towards the end of this phase, the cell passes the G1 restriction point and become committed to subsequent cell division, grows in size, replicates its cytoplasmic mitochondria, and, expresses and stores materials required for DNA replication. This commitment is initiated and strictly controlled through extracellular growth stimulatory signaling. Required genes are transcribed and translated (RNA and protein synthesis) before that.

*S or DNA Synthetic phase:* The Cell replicates its DNA and doubles number of chromosomes into sister chromatids. Each pair of chromosomes replicates simultaneously at a specific time of S phase. At its end, DNA content becomes tetraploid and centriole replicates into two. It takes ~9 hr. We can monitor cell division by observing the progression of the DNA replication.

*G2, Growth or Gap phase 2:* The cell prepares itself for mitosis by checking fidelity of DNA replication and correcting errors. Cell is tetraploid and chromosomes gradually condense into their shortest form. It takes ~4 hr. The required genes are transcribed and translated (RNA and protein synthesis).

*M or mitosis phase:* The cell undergoes actual cell division (breakdown of nuclear envelope, dipolar mitotic spindle formation, prophase, prometaphase, metaphase, anaphase and telophase to segregate sister chromatids) then the cell divides into two daughter cells by cytoplasmic division (cytokinesis). It is the shortest phase ~ 1 hr (20-30 minutes). The act of mitosis can be conveniently divided into four sub-phases: 1) Prophase in which chromosomes are comprised of two sister chromatids joined at the centromere, nuclear membrane breaks down, centrioles move to opposite poles and the spindle forms between them, 2) Metaphase in which chromosomes at their most condensed state and are attach to the spindle by their kinetochores lined up at the equator, 3) Anaphase in which centromeres divide and chromosomes are pulled apart to the two poles by contracting spindle fibers, 4) Telophase in which nuclear membranes for daughter nuclei reform (in yeast the nuclear membrane does not disassembles in principle) and chromosomes de-condense and daughter cells return to interphase.

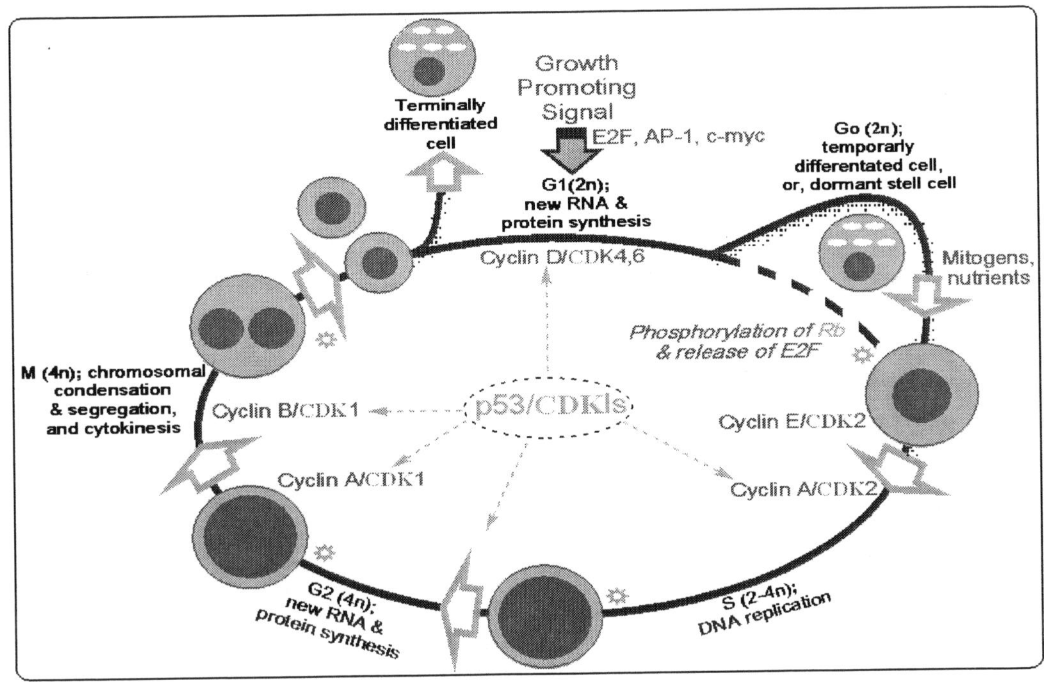


Figure 18. The phases of the cell cycle with the major regulatory events (*; where the checkpoints are).

*Go, No phase* (o = for outside the cycle or time out of the cycle): Normal terminally-differentiated cells are temporarily (e.g., hepatocytes and mucosal crypt epithelium) or permanently (e.g., brain neurons) non-dividing and non-cycling. Starved cells and those in an organ that reached its final size (contact inhibition) will stop at Go. Cell stays dormant until it is instructed by a controlled growth-promoting signal to divide if it is in a temporal stoppage like hepatocyte. Terminally stopping cells has no fate except dying by physiological programmed cell death, i.e., apoptosis. Required genes are transcribed and translated.

*Worth Noting: Analysis of the distribution of cells among the cell cycle phases*

Follow up of the progress of DNA synthesis can be achieved by labeling the newly synthesized DNA by, e.g., $^{3}$H-thymidine and detecting its radioactivity, the nucleotide analog bromodeoxiuridine (BrdU) and its detection by a specific fluorescent anti-BrdU antibody, or by staining DNA in fixed cells with general fluorescent DNA stains such as propidium iodide and monitoring the increase in total DNA content. The later approach is used in flow cytometry to study the proportional distribution of cells among the different cell cycle phases depending on its total DNA content (fluorescent intensity of DNA), i.e., cells with diploid DNA would be at G1, cells with DNA between diploid and tetraploid would be at S phase, and cells with tetraploid would be at G2/M phase. A subG1 phase of lower than diploid DNA content may appear if cells are in apoptosis with fragmented DNA within intact cells (Figure 19).

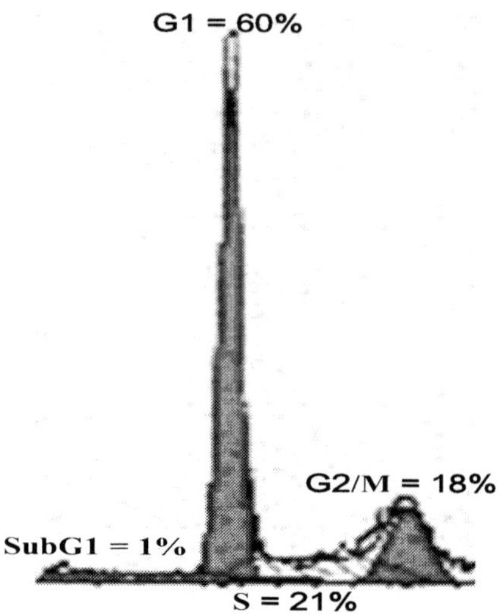


Figure 19. The phases of the cell cycle as sorted by flow cytometry after cell fixation and DNA staining with propidium iodide. The G1 cell portion is the fist red peak to the left, the G2/M portion is the right small red peak, and the blue hatched shoulder between the two peaks is the S phase portion of cells. In case of cells succumbing apoptosis, a preG1 mass of cells (subG1 - the blue very low hill to the left side of the left peak) may show up depending of the nature of treatment before harvesting cells for fixation. The x-axis represents the DNA content and the y-axis represents the number of cells.

## Regulation of the Cell Cycle

The cell cycle phase and progression is highly conserved in evolution with essentially the same molecular machinery. Several extracellular and intracellular regulators control the commitment to cell division. The extracellular regulators include growth factors, cell-cell and cell-matrix attachments. The extracellular effector induces the cell to divide within G1 phase up to but not beyond the restriction point. However, the cell cycle could be temporarily or terminally stopped by intracellular signals at any of the four checkpoints at any phase under extreme conditions. Progress from one phase in the cycle to the next requires completion of events in previous phase and insurance that it was error-free, through extrinsic controlling differential gene expression in each phase and among phases. In any tissue, during embryogenesis and tissue growth or healing cells are distributed among the different phases of the cell cycle in a well coordinated manner. However, cancer cell is an ever dividing cell with cells randomly distributed among cell cycle phases. *In vitro*, the cell cycle phases of a population of normal or cancer cells can be *synchronized* to go as a wave by several approaches including: growth factors and serum deprivation and HMG-CoA reductase inhibitors.

After being committed to cell division at late G1 phase, normal eukaryotic cells become largely unresponsive to extracellular signals and proceed autonomously through DNA replication, G2 and M phases of cell cycle. Before this point, cells could withdraw into the Go quiescent state or succumb to programmed cell death.

The cell cycle has two balanced types of regulators; 1) the stimulatory mitogenic growth factor-immediate and early response genes-Cyclin/cyclin-dependent kinases cascade, and, 2) the inhibitory antimitogenic growth factor-p53/pRb-cyclin-dependent kinase inhibitors cascade.

- *Mitogenic receptor-mediated growth factors signal:*

  *One examples is the* epidermal growth factor, which initiates a cascade of phosphorylation/dephosphorylation that activates, e.g., the Mitogen-Activated Protein Kinases (MAPKs). These kinases phosphorylate and activate *immediate response genes/transcription factors*; AP-1 (c-Jun and c-Fos), c-myc and E2F. The c-myc is implicated in the gradual increase in cell mass during G1 phase. These immediate response genes/transcription factors induce expression of *early response* cyclins and kinases that regulate the catalytic activity of the constitutively active cyclin-dependent proteins kinases *(CDKs)*. CDKs are heterodimers formed of the regulatory cyclin subunit (called so because its level cycles up and down dependent on the cell cycle phase) and the inactive kinase catalytic subunit. Animal cells have at least ten different cyclins (designated A, B, etc.) and at least eight cyclin-dependent kinases (CDK1 - CDK8), which act in various combinations at specific points in the cell cycle. CDKs stimulate the different activities within each cell cycle phase and transitions between cell cycle phases. More than extracellular growth factors, G1-specific kinase activity is also regulated by nutrients, and cell-cell and cell-matrix interactions that are fundamental in regulating cell division through, e.g., contact inhibition of proliferation. CDKs are also involved in the regulation of other functions, such as nuclear envelope breakdown, through phosphorylating and disassembling its lamin C and vimentin proteins.

  The D cyclins are synthesize early in the new cell cycle to activate CDK4 and CDK6 that are also synthesize in an inactive form during G1 in cells undergoing active division. In this stage, the cell cycle regulator retinoblastoma (Rb) protein causes transcription repression through sequestering and inactivating the E2F transcription factor. The D-CDK4/6 as nuclear proteins assembles in a complex form at the late G1 phase. The active complex is a serine/threonine protein kinase that phosphorylates target transcription factors - mainly Rb. The phosphorylated Rb is inactivated and releases an active E2F.

  E2F induces the transcription of specific genes needed in G1 phase and for progression from G1 beyond the restriction point to S phase and DNA replication and metabolism (i.e., *late response genes*) including dihydrofolate reductase, thymidine kinase, histones and DNA replication proteins and the S phase cyclins A and cyclin E, etc). Consequently, gene activation ensues and cell cycle progression to S phase takes place. The S phase cyclin A is initially inactivated because S phase kinase; CDK2 is bound by cyclin E as CDK2/cyclin E complexes that also phosphorylates and inactivates pRb. Through a D-CDK4/6-dependent phosphorylation process; cyclin E is degraded to make the kinase available for cyclin A binding at the G1-S transition. S phase A/CDK2 complex initiates DNA replication by phosphorylating and releasing the inhibitory prereplication MCM complex binding at origins of replication and transit the cell into G2. Cyclin A/CDK1 functions in G2 and transit the cell into M phase. The M phase kinase complex, B/CDK1 (aka, M phase promoting factor, MPF) is produced during G2 phase in an inactive form due to tyrosine/threonine phosphorylation. Towards the end of G2 and early M phases, a phosphatase (CDC25C) dephosphorylates and activates this kinase. The active kinase is rate-limiting for the G2/M transition. Also, the active kinase phosphorylates and activates target proteins necessary for chromosome condensation, nuclear envelope breakdown, spindle formation, alignment of chromosomes and the anaphase promoting complex. The later promotes chromosomal separation and the degradation of CDK1 to enable myosin II-actin binding that initiate the in-vagination of the plasma membrane and cytokinesis.

*To cycle, cyclins level should be controlled by controlling both the rate of synthesis as above and degradation.* Example, cyclins A and B are initially activated then degraded for progression through M phase. Control on degradation is signaled through the nine amino acids sequence near their N-terminal that is called the "destruction box," which targets cyclins A and B for degradation. A protein DBRP (destruction box recognizing protein) recognizes this sequence and brings together the cyclin to be poly-ubiquitinylated and degraded by the protein degradation machinery called proteasome (Figure 20).

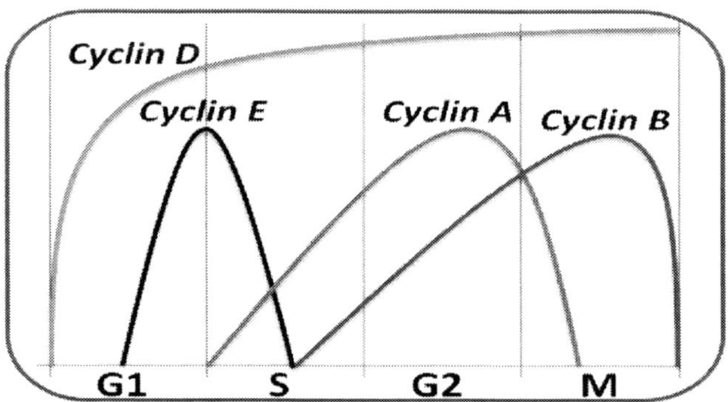


Figure 20. Periodicity of the different cyclin expression and degradation at specific phases of the cell cycle.

- *The network of negative regulators:*

They counteract the above mitogenic cascade of events. Anti-mitogenic growth factors, e.g., Transforming Growth Factor β (TGF-β) signaling activate protein phosphatases to inhibit proliferation. The pRb sequester and inactivate E2F, hence preventing commitment to cell division at G1 phase.

*The cell cycle does not proceed unguarded; there is continuous surveillance of DNA integrity at the 4 restriction checkpoints*, in G1 to check for mutation (*the major check point*); end of S to confirm completion of replication; in G2 to check for new mutation acquired during S phase; and during M phase to check for abnormal spindle formation and abnormal chromosomal segregation and crossovers. In normal cells, the cell cycle will be halt at the phase any of these irreparable DNA defects is detected. This is conducted mainly through extrinsic effectors that are not required in themselves for cell cycle progression. The checkpoint activities are not manifested under conditions in which the potential for errors is minimal (when the frequency of errors is naturally low): only when conditions become stressful and errors are likely to occur do checkpoints become essential survival tools. This criterion formed the basis of early screens to identify mitotic checkpoint components in yeast. Therefore, there are always two ways to progress past a checkpoint. One is to satisfy it by eliminating the condition it monitors. The other is to abrogate the checkpoint itself. These effectors mainly include the transcription factor p53, i.e., the protein with 53 kDa MW. The p53 is highly unstable (normal half-life is 30 minutes) and becomes stable upon its phosphorylation after forming a complex with the damaged DNA through ATM (Ataxia telangiectasia mutated - a serine/threonine protein kinase)/ATR (Ataxia telangiectasia and Rad3 related - a serine/threonine protein kinase) that phosphorylate and activate the checkpoint kinases CHK1 and 2. The stabilized p53 activates transcription of proteins that are CDKs inhibitors (CDKIs), e.g., $p21^{Cip1/Waf1}$, $p16^{Ink4a}$, and $p27^{Kip1}$. CDKIs induce the expression of DNA repair proteins and stop the cell cycle through sequestrating and dephosphorylating immediate response transcription factors-cyclin-CDKs cascade with the help of pRb. If the repair mechanisms fail, p53 *(the guardian of the genome)* induces proteins utilized by the cell to kill itself by apoptosis – the physiological cell death.

## Abnormalities of the Cell Cycle

Abnormalities in cell cycle control due to imbalanced regulation caused by inactivating negative controlling factors and/or activating the positive controlling factors cause serious consequences such as carcinogenesis and embryonic malformations.

- The inactivation of p53 gene by mutations is implicated in most of human cancers. Those individuals who inherit one defective copy of p53 commonly have the Li-Fraumeni cancer syndrome with breast, brain, bone, blood, lung, and skin cancers at an early age. Similarly, pRb gene inactivation by mutation is reported in several types of cancer, particularly, retinoblastoma - the inherited cancer of the retina with increases susceptibility to other cancers in the lung, prostate, and breast. Inactivation of p53 in almost all types of cancer cells and pRb in other cancers allows continuous uncontrolled unchecked cell division without repairing DNA. This leads to worsening of the situation of DNA mutation with time giving rise to more and more aggressive tumors due to acquisition of new mutations. This is why proteins such as p53, pRb and CDKIs are called *tumor suppressor proteins*.
- Infection with oncoviruses and activation of oncogenes by mutation also disrupt the restriction control on cell cycle. These include excessive synthesis of a cyclin or its production at an inappropriate time, e.g., bcl2 oncogene that causes B-cell lymphoma is a cyclin D gene product. Several viral oncoproteins, e.g., E6 and E7 of the human papillomavirus are known to inactivate pRb in skin and uterine cervical cancer. This is why proteins such as growth factors, E2F, E6, cyclins, and CDKs are called *oncoproteins*.

*Worth Noting: Drug inhibitors of the cell cycle*

Immunosuppressants used for organ transplantation patients and for treatment of autoimmune diseases; some antiviral therapy and anticancer chemotherapies utilize chemicals that interfere with cell cycle progression. They inhibit nucleotide and DNA synthesis, cytoskeleton and spindle formation, and cell division and/or induce damage to DNA. They include:

- *Alkylating agents*, e.g., melphalan, cyclophosphamide and bleomycin are potent anticancer drugs because they act at all stages of the cell cycle to cause DNA adducts, cross-linking, mutations and strand breaks.
- *Topoisomerase inhibitors* include two classes; the anthracyclines and the camptothecins. The anthracyclines, e.g., Doxorubicin (from the fungus, *Streptomyces*) inhibit topoisomerase II and induce oxygen free radicals formation and DNA strand breaks. Topoisomerase II inhibitors also include Etoposide one of the plant epipodophyllotoxins. The camptothecins (from the bark of the *Camptotheca accuminata* tree) include irinotecan and topotecan that inhibit topoisomerase I.
- *Disruptors of the microtubules of the mitotic spindle* include the Vinca alkaloids, e.g., vincristine, vinblastine, vinorelbin, and vindesine (from periwinkle plant, *Vinca rosea)* that bind tubulin monomers to inhibit cell cycle. Taxanes (from the Pacific yew tree, *Taxus brevifolia*) including paclitaxel and docetaxel act by abnormally stabilizing microtubules and prevent cell division (cytokinesis).
- *The antimetabolites* are anticancer drugs that interfere with nucleotide metabolism and include two major types: thymidylate synthase inhibitors, e.g., 5-fluorouracil, and, dihydrofolate reductase inhibitors, e.g., methotrexate and trimethoprim.

# DNA Alterations (Mutations and Damage)

Mutation is a detectable alteration of the DNA sequence, whether as small as the alteration of a single base pair, or as gross event as the gain or loss of an entire chromosome. Therefore, mutation could be a gene mutations (sequence alteration); chromosome mutations (structural aberrations in the chromosomes) and/or genome mutations (numerical aberrations in the chromosomes). Mutation reflects impaired repair of DNA and is mostly due to faulty replication (~one mutation/$10^6$ to $10^9$ bp); inherited genomic instability; damaging chemicals or radiation; and abnormal recombination and segregation of the chromosomes. The fate of these alterations include: an altered gene product (function and/or amount); an altered gene regulation, or a change in gene copy number and dose. However, regulated natural recombination is a beneficial cellular process that is restricted to a part of the genome, e.g., exchange of segments of homologous chromosomes during meiosis; transposition of a mobile element within a chromosome or between chromosomes; and gene rearrangement, e.g., for the immunoglobulin genes.

Germ cell mutation is vertically heritable, whereas somatic cell mutation affects daughter cells within the organism, is horizontal transmission and is not heritable. Inborn errors are due to heritable mutations, whereas other diseases like cancer and aging are due to vertically and horizontally acquired mutations.

However, mutation through altering the DNA sequence was vital to our understanding of gene function, help isolating genes and to assign specific functions to new encoded proteins, e.g., the chloride channel cystic fibrosis trans-membrane conductance regulator (CFTR) or to a specific regions of a protein through specific site-directed mutagenesis. The later not only helped understanding DNA and protein functions but also helped generating neo-proteins with super-functions and altered specificities.

## Factors that Increase the Natural Rate of Mutation

a. *Spontaneous errors of DNA replication* that skip the proofreading function of the DNA polymerases. The error rate for DNA synthesis in *E. coli* is 1 in $10^6$-$10^9$ but its real genome rate is reduced to only 1 in $10^{10}$ - $10^{11}$ due to corrections by the mismatch repair system. Replicative forward or backward *slippage* most frequently occur at DNA regions with short repeated sequences that leads to duplicated replication of a sequence or skipping another. This is the major cause of expansion of microsatellite sequences that increases their existing population allelic forms and related diseases. But these diseases involve also deficiency in the DNA repair proteins.
b. *Inherited genomic instability and defective repair.*
c. *Physical mutagens* (most common) such as UV, X and $\gamma$ radiations. The visible light, in presence of photosensitizers, such as porphyrins and toxins, is able to produce mutagenic free radicals. High frequency electromagnetic waves may predispose to mutation and cancer.
d. *Chemical mutagens* such as anticancer base analogs and alkylating agents.
e. *Environmental pollutants* radiation and chemicals such as nitrous acid, oxidative free radical generators and alkylating agents, e.g., methyl halides - products of tobacco smoking and incomplete ignition of fuel.
f. *Viral infections* particularly with oncogenic retro- and adeno-viruses.

*Worth Noting: Mutagen, carcinogen and teratogen*

*It is important to discriminate between, a mutagen that causes mutation; a carcinogen that is a mutagen that causes cancer through mutation; and a teratogen that is a mutagen or non-mutagen chemical (e.g., vitamin A and retinoic acid) that causes embryonic developmental abnormalities.* The mutagenic potential of a chemical can be tested by *Ames test* developed by Bruce Ames. The test uses a mutant bacterium, *Salmonella typhimurium* strain that cannot grow in the absence of histidine due to mutation in the genes of histidine biosynthesizing enzymes. Thus, this strain is called $His^-$ phenotype. Growing the bacterium in a histidine-free medium that contains the potential mutagen could acquire the bacterium new mutations that reverse the original histidine mutations. This would enable some bacterial cells to synthesize histidine and acquire $His^+$ phenotype and grow into colonies in such conditions. The natural form of some chemical, e.g., benzo(a)pyrene, is not mutagenic but become so after activating metabolism during drug detoxification in liver or kidney. To test the mutagenicity of such chemicals (*procarcinogen*) requires their incubation first with homogenates of such tissues and the extracted metabolites are then incubated as above in the

Ames test. Mutagenicity of a chemical in the Ames test correlates positively with carcinogenicity test of the same chemical in animals. Among chemicals known to be mutagenic in bacteria, 85% are carcinogenic (cancer-causing) in animals.

## Pattern of DNA Mutations

DNA mutations can be divided into five categories depending upon the number of nucleotides involved or the type of defect.

a) *Single base alteration (point mutation)* that is due to:
   1. *Deamination.* Adenine deaminates into hypoxanthine that base-pairs with C and cytosine deaminates into uracil that base-pairs with A. There is a basal rate of spontaneous deamination that is mainly hydrolytic. Mutagens, e.g., nitrous acid ($HNO_2$) and sodium bisulfite are deaminating agents; that is mainly oxidative deamination.
   2. *Depurination or depyrimidination.* It is the removal of a purine or a pyrimidine base from the nucleotide leaving an exposed deoxyribose attached to the sugar-phosphate backbone. Spontaneous depurination occur at 37 $^{o}$C at rate of 5000-10,000/cell/day due to the thermal liability of the purine N-glycosidic bond. It is due to; hydrolysis, thermal disruption, or, carcinogens such as the fungal aflatoxin.
   3. *Alkylation and adducts.* It affects mainly guanine through covalent addition of an alkyl radical (e.g. $-CH_3$). Methyl halides, ethylmethane sulfinate and dimethylnitrosamine are example alkylating agents. Benzo(a)pyrene and aflatoxin B1 are metabolically activated by cytochrome P450 into epoxides reactive with the N7- of guanosine that cause GC-AT conversion.
   4. *Oxidation.* Oxidative conversion of deoxyguanosine into 8-hydrox-deoxyguanosine that rearranges into 8-oxo-deoxyguanosine. The later base pairs with A leading to GC–AT conversion.
   5. *Insertion or deletion of a nucleotide.* Flat intercalating mutagens, e.g., ethidium bromide and acridines intercalate between base-pairs and cause insertion or deletion mutations that mostly lead to frame-shift mutation.
   6. *Incorporation of base-analog*, e.g., 5-bromouracil (a thymine analog) and 2-aminopurine. They are more likely to be incorporated during replication than normal nucleotides. They induce AT- GC conversion because they preferentially exist in the lactim form.

b) *Two bases alteration* that is due to:
   1. UV-induced thymine–thymine dimer that induces later deletion mutation.
   2. Two nucleotides cross linkage in same DNA strand caused by bi-functional alkylating agent.

c) *Strand breakage: Single-strand breaks* are less harmful than *double strand breaks* in the phosphodiester backbone. The strand breaks are generally caused by:
   1. Ionizing radiation that has direct and indirect damaging effect through free radicals.

2. Radioactive disintegration of incorporated element(s).
3. Oxidative free radical generating reactions and chemicals.
4. Very high doses of alkylating carcinogens such as nitrogen mustards.
5. DNA-strand breaking chemotherapeutic agents, e.g., cisplatin.

d) *Cross-linkage:*
1. Two bases in same strand or opposite strands.
2. DNA cross-linkage to histones or other proteins.

e) Structural and Numerical Chromosomal Abnormalities.

## Structural and Numerical Chromosomal Abnormalities

They are caused by faulty segregation; recombination or gross damage. Chromosome abnormalities are changes resulting in a visible alteration of the chromosomes. How much can be seen depends on the technique used. The smallest loss or gain of material visible by traditional cytogenetic methods on standard preparations is about 4 megabases of DNA. However, molecular cytogenetics using *fluorescence in situ hybridization (FISH)* and *PCR* allow much smaller changes to be seen. Thus, molecular cytogenetics has removed any clear dividing line between changes described as chromosomal abnormalities and changes described as molecular DNA defects. Therefore, chromosomal abnormality could alternatively be defined as an abnormality produced by specifically chromosomal mechanisms such as mis-repair of broken chromosomes due to improper recombination or by malsegregation of chromosomes during mitosis or meiosis or due to abnormal paternal origin of chromosomes. A couple should not be sorry for an early abortion - along traumatic, nutritional, toxic and infective causes were excluded, because abortion is most probably due to unwanted chromosomal abnormality. This happens most frequently with young ladies with large a ova reserve. Such abortions get less frequent with the depletion of the ova reserve with a bigger chance to have babies afflicted at the women older productive ages. **Figure 21** illustrates prevalence and types of chromosomal abnormalities and their

About 20% of conceptions have some sort of chromosomal disorder - the most damaging of which are changes in the chromosome number - but because of the lethal effects of such disorders, the number actually born is only about 0.6%. A chromosomal abnormality may be present in all cells of the body (constitutional abnormality), or may be present in only certain cells or tissues (somatic or acquired abnormality). Constitutional abnormalities must have been present very early in development, most likely the result of an abnormal sperm or egg, or maybe abnormal fertilization or an abnormal event in the early embryo. By contrast, an individual with a somatic abnormality is a *mosaic*, i.e., containing cells with two different chromosome karyotypes, with both cell types deriving from the same zygote. Chromosomal abnormalities, whether constitutional or somatic, mostly fall into one of three categories:

- Numerical abnormalities.
- Structural abnormalities.
- Wrong parental origin-induced chromosomal complements abnormalities.

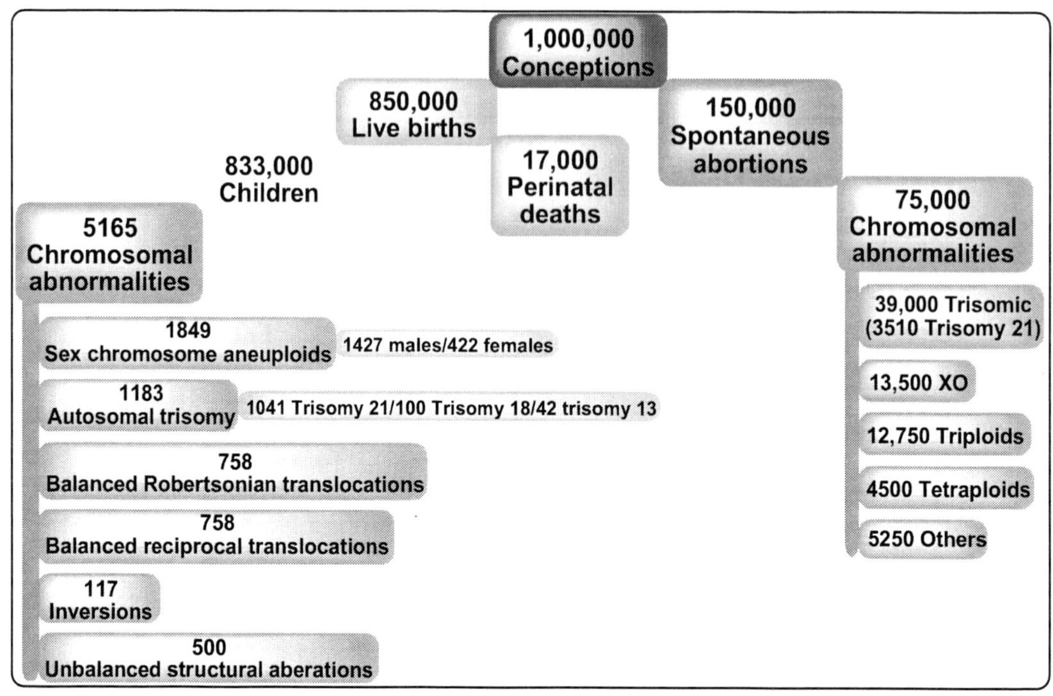


Figure 21. The fate and types of chromosomal abnormalities in a million implanted human zygotes.

## The Numerical Chromosomal Abnormalities

Numerical abnormalities involve the loss and/or gain of a whole chromosome or chromosomes including both autosomes and sex chromosomes. Generally chromosome loss has a greater effect on an individual than does chromosome gain although these can also have severe consequences. Another general rule is that loss or gain of an autosome has more severe consequences than loss or gain of a sex chromosome. Occasionally an individual carries an extra chromosome which cannot be identified by its banding pattern, these are called *marker* chromosomes. The introduction of FISH techniques has been a valuable tool in the identification of marker chromosomes. *There are three classes of numerical chromosomal abnormalities: polyploidy, aneuploidy and mixoploidy.*

*Euploidy (Polyploidy)* means having complete chromosome sets (1*n*, 2*n*, 3*n*, 4*n* etc.). It is a numerical chromosomal abnormality which involves the addition or loss of a complete set of chromosomes (*n*). Between 1 and 3% of recognized human pregnancies are *triploid, 3n* (69 total). The most usual cause is two sperm fertilizing a single egg (*dispermy*, two thirds of cases) but it also could be due to participation of a diploid gamete (Figure 22). Triploids are life-incompatible and very seldom survive to term. *Tetraploidy* is much rarer, life-incompatible and is usually due to failure to complete the first zygotic division after replicating DNA into 4*n* (92 total). Although constitutional (whole body) polyploidy is rare and incompatible with development, all normal people have some polyploid cells.

*Aneuploidy* is a numerical chromosome change which do not involve a whole set of chromosomes. It is usually the consequence of a failure of a single chromosome (or bivalent) to complete division. Aneuploidy causes several specific genetic syndromes. Cancer cells often show extreme aneuploidy, with multiple chromosomal abnormalities. *Aneuploid cells arise through two main mechanisms; nondisjunction and anaphase lag. Nondisjunction* is the failure of paired chromosomes to separate (disjoin) in anaphase of meiosis I, or failure of sister chromatids to disjoin at either meiosis II or at mitosis (Figure 23). Non-disjunction in meiosis produces gametes with 22 or 24 chromosomes, which after fertilization by a normal 23 chromosomes gamete results in a monosomic (45 momosomy) or a trisomic (47 polysomy) zygote, respectively. Non-disjunction in mitosis produces a mosaic. *Anaphase lag* is the failure of a chromosome or chromatid to be incorporated into one of the daughter nuclei following cell division, as a result of delayed movement (lagging) during anaphase. Chromosomes that do not enter a daughter cell nucleus are lost.

*Clinical examples of aneuploidy* include; trisomies [e.g., chromosome 21 (Down's syndrome), 13, 18 (Edward's syndrome), and sex chromosome aneuploidies] and monosomies, e.g., Turner's syndrome (45, X0). *Trisomy* is presence of three copies of a particular chromosome in an otherwise diploid cell, when a gamete with more than 23 chromosomes participates in conception. It involves autosomes such as 8, 13, 18 or 21 and sex chromosomes X or Y. *Down's syndrome* (47, XX or XY, +21) that is a trisomy for chromosome 21. Down's syndrome incidence at birth sharply increases with advancement of mother age particularly after 35 Y (1:1500 at age 20 and 1:30 above age 45). It could be due to aging effect on oocytes and exposure to damaging agents such as drugs, chemicals and radiations. However, 15% of the extra chromosome comes from the father. Most cases arise from non-disjunction in the first meiotic division. A small proportion of cases are mosaic and these probably arise from a non-disjunction event in an early zygotic division. About 4% of cases arise by inheritance from a parent who is a carrier of a balanced chromosome translocation. Down syndrome is responsible for about 1/3 of all cases of moderate to severe mental handicap (IQ of less than 50). The patients suffer from small square head, upward sloping of the eye, small and malformed ear, an open mouth with large protruding tongue, short and study hand with fingers curled inwards and the palmer crease is single. They have 20 times greater chance to develop acute leukemia, congenital heart defects, and mental retardation.

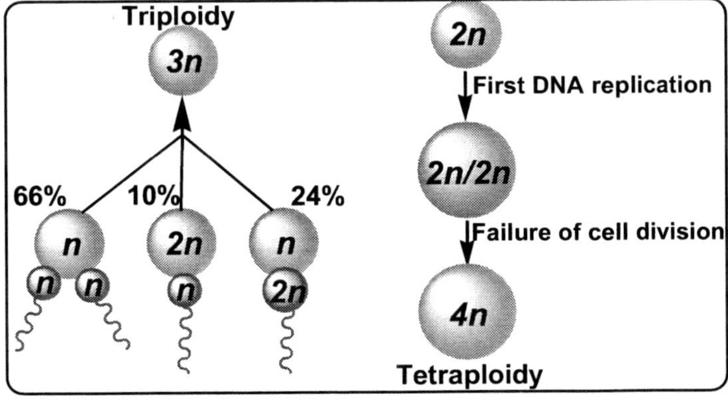


Figure 22. Possible causes of euploidy.

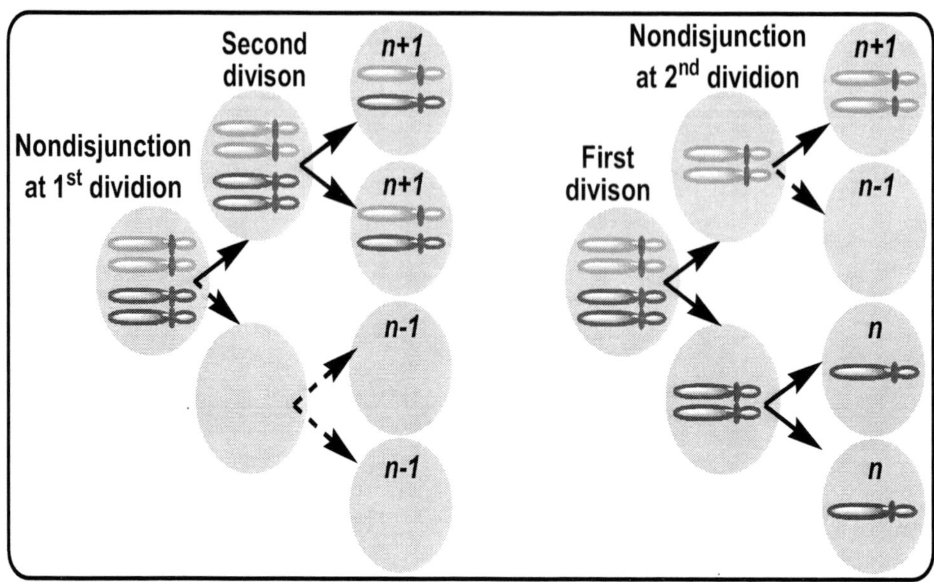


Figure 23. Non-disjunction as the cause of aneuploidy.

*Trisomy 13* has an incidence of ~1 in 5000 live births. 50% of these babies die within the first month and very few survive beyond the first year. There are multiple dysmorphic features. Most cases, as in Down's syndrome, involve maternal non-disjunction. Again, a significant fraction has a parent who is a translocation carrier. *Trisomy 18 (Edward's syndrome)* has an incidence of ~1 in 3000. Again most babies die in the first year and many within the first month.

*Sex chromosome aneuploidies* are far milder and hence common in live-births than those involving autosomes because of X chromosome inactivating imprinting and because of the rarity of genes on the Y chromosome. *47, XYY males* have an incidence of 1 in 1000 male births. It may be without any symptoms. Patients are tall but normally proportioned and have 10 - 15 points reduction in IQ compared to sibs. Strangely, although they are fertile they do not seem to transmit either this condition or Klinefelter's syndrome. *47, XXX females* have an incidence among women of ~1 in 1000 with an extra X chromosome. It seems to do little harm, individuals are fertile and do not transmit the extra chromosome. They do have a reduction in IQ comparable to that of Klinefelter's males. *Klinefelter's syndrome,* with polysomy X (47, XXY) males. The incidence at birth is about 1 in 1000 males. Testes are small and fail to produce normal levels of testosterone leading to infertility and breast growth (gynaecomastia) in about 40% of cases and to poorly develop secondary sexual characteristics. There is no spermatogenesis. These males are taller and thinner than average due to delayed puberty and may have a slight reduction in IQ. Very rarely more extreme and severe forms of Klinefelter's syndrome occurs where the patient has 48, XXXY or even 49, XXXXY karyotype. Testosterone replacement may improve the case.

*Monosomy* is the corresponding lack of a chromosome out of the two homolog pair. Autosomal monosomy is life-incompatible and leads to early abortion, whereas, monosomy of chromosome X is less severe. *Turner's syndrome (45, X0) females* have an incidence of ~1 in 500 female births. However, 99% of Turner syndrome embryos are spontaneously aborted.

Individuals are very short, they are usually infertile. Characteristic body shape changes include a broad chest with widely spaced nipples and may include a webbed neck; gonadal agenesis (absence of ovaries), no secondary sex characteristics, short stature, non-pitting edema in the hand and feet and congenital heart defects. IQ and lifespan are unaffected. Mosaicism with normal karyotype makes the condition less severe. Estrogen replacement enhances secondary sex characteristics and may increase the stature, but does not reverse infertility.

*Mixoploidy* includes *mosaicism* (an individual with two or more genetically different cell lines all derived from a single zygote) and *chimerism* (an individual with two or more genetically different cell lines originating from different zygotes). The abnormalities that are lethal in their constitutional form may be life compatible in mosaics with normal karyotype. *Aneuploidy mosaics* are common. For example, mosaicism resulting in a proportion of normal cells and a proportion of aneuploid cells (e.g., trisomic) can be ascribed to non-disjunction or chromosome lag occurring in one of the mitotic divisions of the early embryo (the monosomic cells that are formed usually die out). *Polyploidy mosaics* (e.g., human diploid/triploid mosaics) are occasionally found. As gain or loss of a haploid set of chromosomes by mitotic non-disjunction is most unlikely, human diploid/triploid mosaics most probably arise by fusion of the second polar body with one of the cleavage nuclei of a normal diploid zygote.

## The Structural Chromosomal Abnormalities

They result from mis-repair of chromosome(s) breaks or from malfunction of the recombination system followed by an abnormal rearrangement, deletion or translocation. Chromosome breaks occur either as a result of damage to DNA induced by ionizing radiation, e.g., X rays, viral infections, chemical pollutants and extreme changes in cellular environment or as part of the mechanism of recombination. More than one double strand DNA breaks could become rejoined in the wrong combinations leading to a spectrum of possibilities. Its consequences depend on the amount of genetic material lost, where, massively damaged cells will die. In G2 phase of the cell cycle chromosomes consist of two chromatids. Breaks occurring at this stage are manifest as chromatid breaks, affecting only one of the two sister chromatids. Breaks occurring in G1 phase, if not repaired before S phase, will affect both chromatids.

Any resulting chromosome that has no centromere (acentric) or two centromeres (dicentric) will not segregate stably in mitosis, and will eventually be lost. Chromosomes with a single centromere can be stably propagated through successive rounds of mitosis, even if they are structurally abnormal. Meiotic recombination between mis-paired chromosomes is a common cause of translocations, especially during spermatogenesis. Structural chromosomal rearrangements are *balanced* if there is no visible net gain or loss of chromosomal material, and *unbalanced* if there is net gain or loss. The origin of the different chromosomal rearrangements is depicted in Table 4.

**Table 4. The origin of the different chromosomal rearrangements**

| Breaking and Rejoining | Crossover and recombination at repetitive sequences |
|---|---|
| 2 3; 1 2 3 4; lost; 1 4; Deletion | 3; 4; 2; 1; Repetitive sequences; 1 2 3 4; 1 Deletion 4; 3 2 lost |
| Deletion; 1 2 4; 1 2 4; 1 2 3 3 4; 1 2 3 3 4; Duplication | 1 2 3 4; 1 2 3 4; 1 Deletion 4; Duplication; 1 2 3 2 3 4 |
| 1 2 3 4; Inversion; 1 2 3 4; 1 3 2 4 | 2 3; 1 4; 1 2 3 4; 1 3 2 4; Inversion |
| 1 2 3 4; 5 6 7 8 9 10; 1 2 8 9 10; 5 6 7 3 4; Reciprocal translocation | 1 2 3 4; 5 6 7 8 9 10; 1 2 8 9 10; 5 6 7 3 4; Reciprocal translocation |

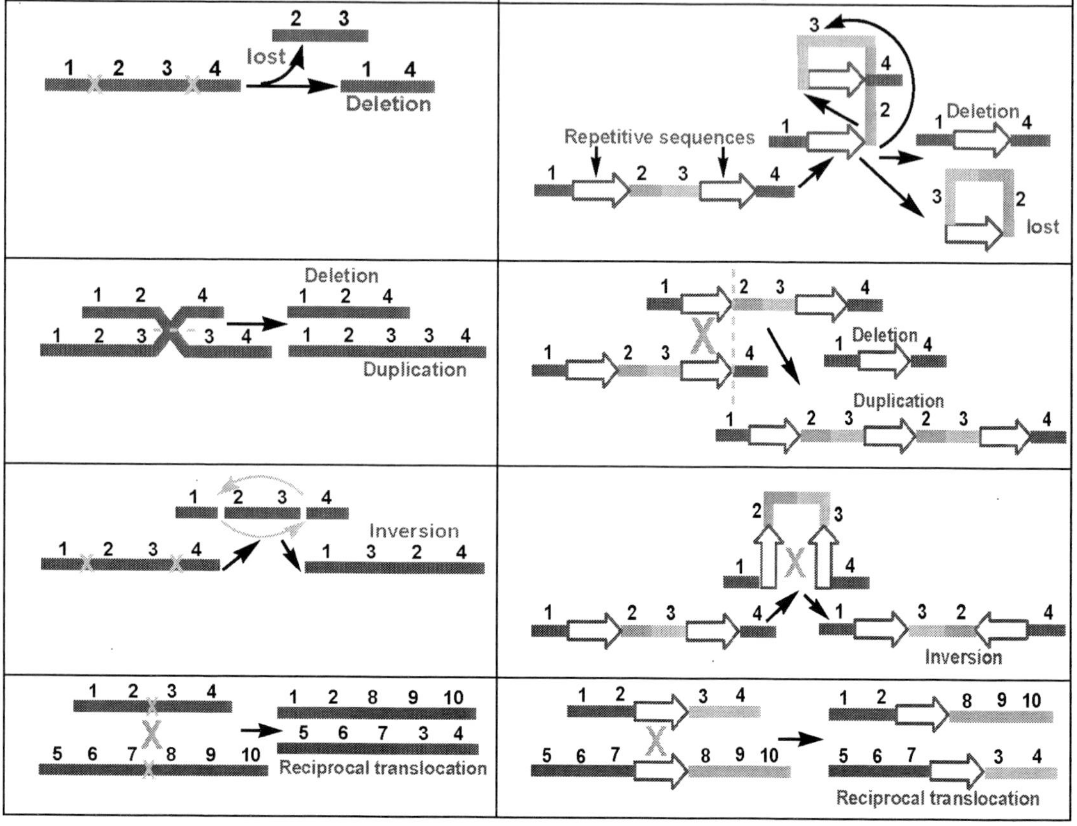

*Balanced rearrangements* include peri- and para-centric inversions, and balanced translocations. They have no effect on the phenotype, although there are important exceptions to this:

- A chromosome break may disrupt an important gene;
- The expression of a gene even with undisrupted coding sequence may be affected by, e.g., its separation from a regulatory element, or its moving into an inappropriate chromatin environment, e.g., trans-locating an active gene into heterochromatin;
- Balanced X-autosome translocations cause problems with X-inactivation.

*Peri- and para-centric inversions* are one of the more common structural rearrangements that occur when there are two breaks within a single chromosome and the broken segment flips 180° (inverts) and reattaches to form a chromosome that is structurally out-of-sequence. If this includes the centromere then the inversion is termed pericentric. If it excludes the centromere then it is a *paracentric inversion*; (Figure 24). The two have slightly different genetic consequences. 1% of the UK population is heterozygous for a pericentric inversion of chromosome 9. This is absolutely without genetic consequences. There is usually no risk for problems to an individual if the inversion is of *familial* origin (i.e., inherited). There is a slightly increased risk if it is a *de novo* (new) mutation due possibly to an interruption of a key gene sequence. At meiosis heterozygous inverted chromosomes have difficulty pairing and can only do so by the formation of a loop. If a recombination event occurs within the inverted loop the consequence will be duplication and a deletion. If the inversion is paracentric then the centromere itself may be duplicated (which gives rise to a dicentric fragment which will try to go to both poles at anaphase I with dire consequences) or deleted (giving rise to an acentric fragment which gets left behind on the metaphase plate).

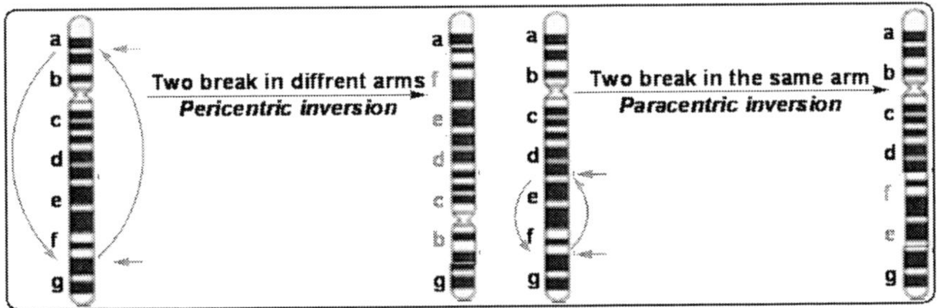


Figure 24. Peri- and para-centric inversions.

*Balanced translocations* do not cause net gain or loss of chromosomal material, two chromosomes have been broken and rejoined in the wrong combination. Figure 25 shows a reciprocal translocation between the imaginary chromosomes "M" and "N". Balanced reciprocal translocation is unlikely to have any severe consequence for the cell because, even if one of the breakpoints lies within a gene, most mutations are recessive. It is possible that an oncogene may be activated by the translocation and this can lead to cancer. The classic example is the *chromosome 9/chromosome 22 reciprocal translocation in chronic myeloid leukemia*. Translocations can however give rise to difficulties with reproduction. During the first meiotic prophase, the chromosomes align in pairs to form bivalents. However, a heterozygote for a reciprocal translocation forms instead a 'tetravalent'. The chromosomes will segregate in the first meiotic division. Many possibilities are open, only one, the one shown in the figure, leads to balanced gametes. This is known as "alternate segregation," where the two intact chromosomes must both move to one pole and the two translocation chromosomes must both move to the other.

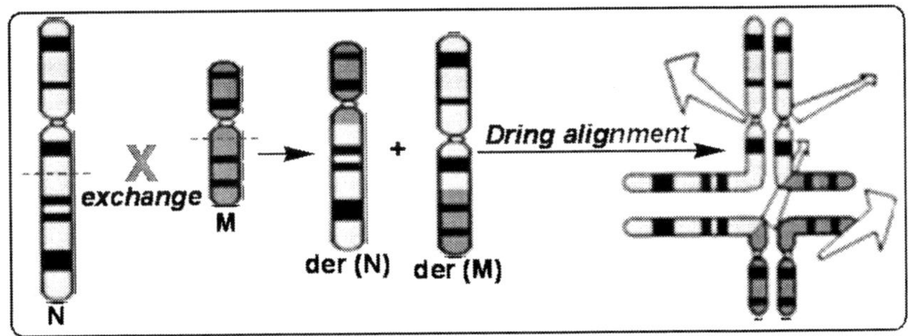


Figure 25. Reciprocal translocation between the imaginary chromosomes "M" and "N"; der = derivative chromosome and the nature of their subsequent alignment with two normal chromosomes.

*Unbalanced rearrangements* can arise directly, through deletion or, rarely, duplication, unbalanced translocations, isochromosomes, ring chromosomes, or indirectly by malsegregation of chromosomes during meiosis in a carrier of a balanced abnormality. *Deletions* may either be interstitial (Figure 26) or terminal. If big enough to be visible, a deletion will remove many genes and will probably give rise to a severe phenotype. *Interstitial deletion* of 15q- removes many genes that cause; Angelman (46, XY, del(15)(q12)(mat)), Prader-Wili (46, XX, del(15)(q12)(pat)) or WAGR (Wilms tumour, Aniridia, ambiguous Genitalia, and mental Retardation) syndromes.

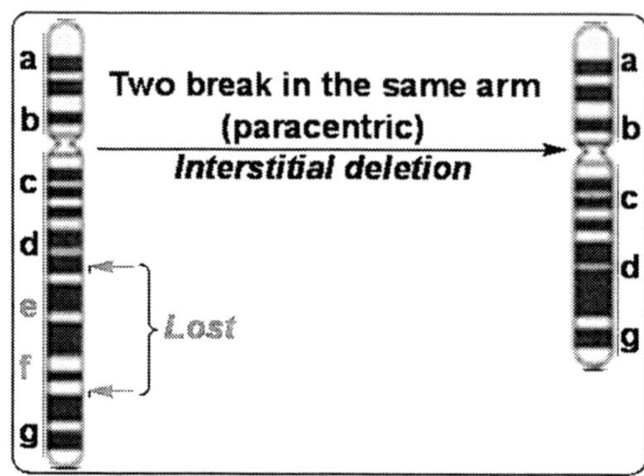


Figure 26. Interstitial deletion.

*Terminal deletions* have only one breakpoint, they extend to the telomere. For example the karyotype shown below is of a baby with Cri du chat syndrome in which a small part of the distal region of the short arm of chromosome 5 is deleted. Babies with this syndrome have a combination of symptoms which include pinched facial features, mental retardation and developmental delay. The characteristic feature, for which the syndrome is named, is a "mewing" cry. The characteristic cry may be separated genetically from the facial dysmorphology and developmental delay, since a small terminal deletion may have the cry only, whereas, a larger deletion extending further towards the centromere will include the genes whose hemizygosity is responsible for the other symptoms.

*Unbalanced translocations* may arise spontaneously and are also likely to arise as an offspring of a balanced carrier. Their exact nature will be unpredictable but are likely to have severe symptoms. Such translocations are extremely useful research-wise by pinpointing genes responsible for the conditions expressed by their carriers. *Robertsonian translocations* are a special case of 'almost balanced' translocations. Robertsonian translocations involve any two out of the acrocentric chromosomes; 13, 14, 15, 21 and 22, where, the centromere is very close to one end. The short arms (antennae) contain few, if any, genes except for many tandemly-repeated copies of the ribosomal RNA genes. Every diploid cell thus contains 10 copies of the block of repeated genes. A Robertsonian translocation is a fusion between the centromeres of two of these chromosomes with loss of the short arms. This results in a chromosome with two long arms, one derived from each chromosome (Figure 27). As with balanced reciprocal translocation carriers, there will be difficulties at meiosis in ensuring the correct segregation of the chromosomes to make a balanced set with Robertsonian translocation carriers. Chromosomes pair in meiotic prophase to form a trivalent and balanced gametes will only be formed when the translocation chromosome goes to one pole and both of the normal chromosomes to the other pole. Robertsonian carriers therefore suffer a similar reduction to their fertility as do carriers of reciprocal translocations and couples in which one of the partners has a translocation may have a number of early spontaneous abortions. Robertsonian translocations involving two chromosomes 21 possess a special problem of their own; one of the possible unbalanced gametes will contain effectively two copies of chromosome 21 (when the translocation chromosome and the normal chromosome 21 segregate to the same pole). They are thus at risk of producing a baby with *Down's syndrome*.

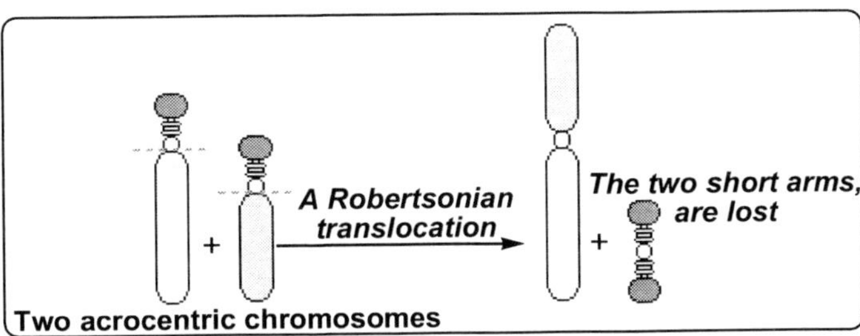


Figure 27. The Robertsonian translocation.

*Isochromosomes* are produced from the rare splitting of a particular chromosome's chromatids "the wrong way" in mitosis (or meiosis II) to generate two symmetrical chromosomes one containing both long arms attached and moves to one pole, and the other contains both short arms attached and moves to the other pole (Figure 28). The consequence is the formation of two isochromosomes. These are simultaneously duplicated for the genes in the retained arm and deleted for the genes in the other. They are believed to arise from an abnormal U-type exchange between the sister-chromatids just next to the centromere of a chromosome. The prognosis is poor except for iXq (isochromosome of the long arm of the X). Isochromosomes are rare except for i(Xq) and i(21q) as an occasional cause of *Down's syndrome*.

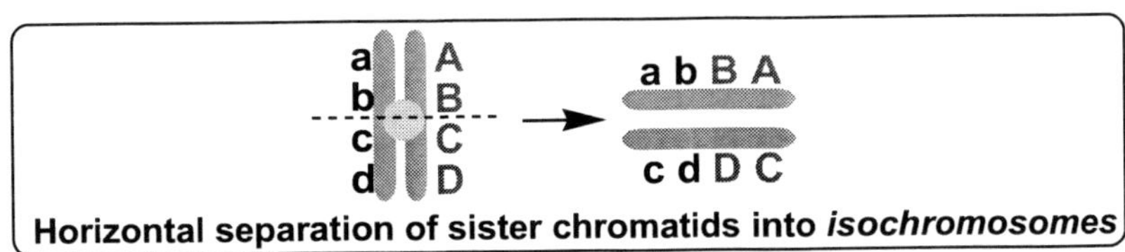


Figure 28. Isochromosome formation.

*Ring chromosomes* are produced when a mutation event removes both telomeres and the repair system seals the ends together forming a ring chromosome. This may have deletion for subtelomeric genes at both ends of the chromosome (Figure 29). The symptoms will depend on the extent of the deletion. Surprisingly, ring chromosomes are mitotically stable. One might expect them to get hopelessly entangled during DNA replication and at the very least to be concatenated when the time comes to separate into the two poles. While this undoubtedly happens, it is not a frequent occurrence.

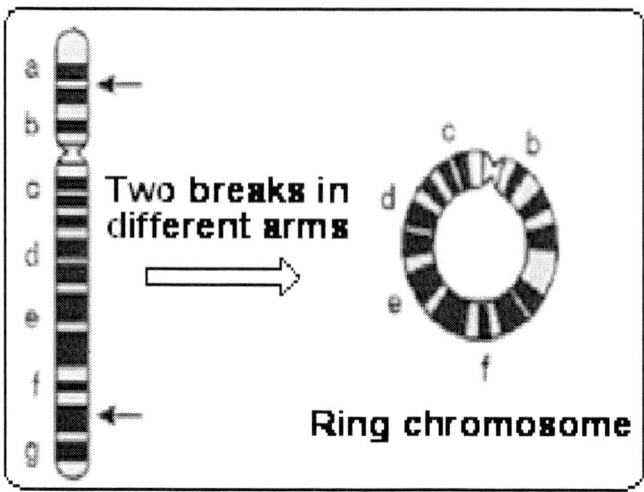


Figure 29. The ring chromosome formation.

## Abnormalities in the Chromosomal Complements due to Wrong parental Origin

It is not enough to have the correct number and structure of chromosomes; but must also have the correct parental origin. 46, XX conceptuses in which both genomes originate from the same parent (*uniparental diploidy*) never develop correctly. For some individual chromosomes, having both homologs derived from the same parent (*uniparental disomy*) also causes abnormality. A small number of genes are received imprinted as their parental origin. It is assumed that the abnormalities of uniparental disomy and uniparental diploidy are caused by abnormal expression of such imprinted genes.

*Uniparental diploidy* is seen in hydatidiform moles, abnormal conceptuses with a 46, XX karyotype of exclusively paternal origin. Molar pregnancies show widespread hyperplasia of the trophoblast but no fetal parts, and have a significant risk of transformation into choriocarcinoma. Genetic marker studies show that most moles are homozygous at all loci, indicating that they arose by chromosome doubling from a single sperm. Ovarian teratomas are the result of maternal uniparental diploidy. These rare benign tumors of the ovary consist of disorganized embryonic tissues, without extra-embryonic membranes. They arise by activation of an unovulated oocyte.

*Uniparental disomy* can be *isodisomy*, where both homologs are identical, or *heterodisomy*, where they are derived from both homologs in one parent. The usual cause is thought to be trisomy rescue. A conceptus that is trisomic and would otherwise die occasionally loses one chromosome by mitotic nondisjunction or anaphase lag from a totipotent cell. The euploid progeny of this cell form the embryo, while all the aneuploid cells die. If each of the three copies has an equal chance of being lost, there will be a two in three chance of a single chromosome loss leading to the normal chromosome constitution and a one in three chance of uniparental disomy (either paternal or maternal). Uniparental isodisomy may possibly arise by selection pressure on a monosomic embryo to achieve euploidy by selective duplication of the monosomic chromosome. Rare cases of *cystic fibrosis* (a common autosomal recessive disease) have occurred in which one parent was a heterozygous carrier of the disease but the second parent had two wild type alleles. The child had received two copies of the mutant chromosome 7 from the carrier parent and no chromosome 7 from the unaffected parent.

## Sampling and the Basic Techniques for Conventional and Molecular Chromosomal Cytogentics

Obtaining dividing cells to see the chromosomes directly from the human body is difficult. A variety of tissue types can be used to obtain chromosome preparations. Bone marrow is a possible source. However, it is much easier to take an accessible source of non-dividing cells and culture them in the laboratory; e.g., peripheral blood that is the material of choice. Other common sources include fibroblasts grown from skin biopsies. *Prenatal chromosomal analysis* is indicated in cases that include; late maternal age; family history; and past history of early spontaneous abortions. At about 12 - 14 weeks of gestation it is possible to obtain fetal cells by amniocentesis. In this technique, a needle is inserted into the amniotic sac and fluid is drawn off and fetal cells (amniocytes) are cultured *in vitro* and their metaphase chromosomes are examined. At 9-10 weeks of gestation, chorionic villus biopsy (a small amount of the placental tissue) is recovered through the cervix and cultured *in vitro* for chromosomal preparation. A single, totipotent cell from an 8-cell stage of *in vitro* fertilization embryo can be examined directly and the information can be used to decide on which embryo to reintroduce into the mother/foster mother uterus.

Most chromosome abnormalities originate during meiosis that necessitates studying human gametes. *Meiosis* can only be studied in testicular - including the sperms, or ovarian samples. Female meiosis is especially difficult, as it is active only in fetal ovaries, whereas male meiosis can be studied in a testicular biopsy from any post-pubertal male. The results of meiosis can be studied by analyzing chromosomes from sperm – since sperms may show abnormalities in otherwise normal persons; although the methodology for this is cumbersome. Meiotic analysis is used for some investigations of male infertility. Techniques for human sperm karyotyping include; the hamster oocyte system to reactivate human sperm allowing the analysis of the pronuclear chromosomes, Fluorescent In Situ Hybridization (FISH), single sperm polymerase chain reaction (PCR) analysis, and immunocytochemical labeling of the synaptonemal complex (the proteinaceous structure linking homologous chromosomes in prophase of meiosis 1 observed in testicular biopsy cells). Moreover, antibodies against the DNA mismatch repair protein MLH1 (see alter in the course) identify the sites of meiotic exchange.

In the basic processing technique, cells, e.g., blood T lymphocytes, are induced to divide by treatment with lectins such as phytohemagglutinin and are allowed to grow for 48-72 hours. Nevertheless, because M phase occupies only a small part of the cell cycle, few cells will be actually dividing at any one time. The mitotic index (proportion of cells in mitosis) is increased by treating the culture with a spindle disrupting agent such as colcemid to arrest cells at metaphase. Often it is preferable to study prometaphase chromosomes, which are less compacted and have a higher banding resolution. Cell cultures are first synchronized by depraving cells from thymidine or serum. By trial and error, the time after release can be determined when a good proportion of cells are in the desired prometaphase stage. Cells are harvested and treated by hypotonic solution to swell them followed cell fixation. This allowed the first good quality preparations to be made in 1956.

Various pre-staining treatments involving denaturation and/or proteolytic enzyme (e.g., trypsin) digestion, followed by incorporation of a DNA-specific dye, help increasing the resolution in the stained preparation. This allows for the detection of subtle changes in chromosome structure. The most common staining treatment is called G-banding using Giemmsa`s staining. A variety of other staining techniques are available to help identify specific abnormalities. Depending on the nature of the stain and banding pattern required, cells could be denatured by mild severe heat treatment in saline or denatured in saturated solution of barium hydroxide. Typically 15-20 cells are scanned and counted with at least 5 cells being fully analyzed. During a full analysis each chromosome is critically compared band-for-band with its homolog. It is necessary to examine this many cells in order to detect clinically significant mosaicism. Typical high-resolution banding procedures for human chromosomes can resolve a total of 400, 550 (Figure 30) or 650 bands.

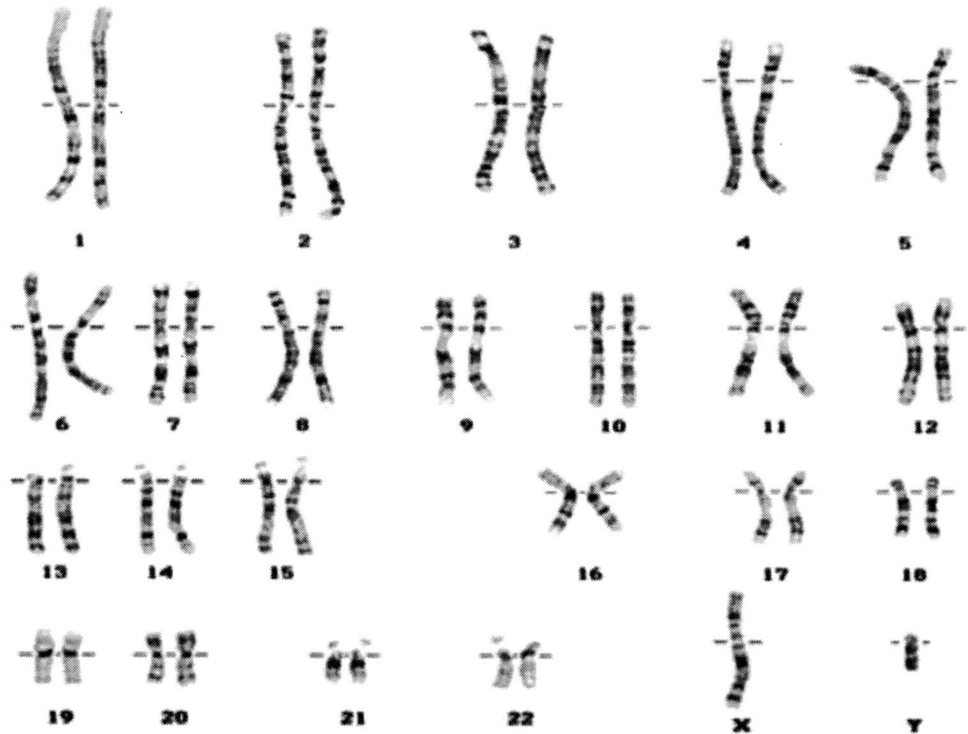


Figure 30. A typical, normal male karyotype at the 550 bands resolution level. The homologous pairs of chromosomes are arranged according to size, banding pattern, and centromere location. Each chromosome is compared band-for-band with its homolog for any structural differences. This male karyotype is: 46, XY (i.e., 46: a normal total number of chromosomes, and XY: a normal set of sex chromosomes.

The ability of a single stranded DNA molecule to bind specifically to its complementary strand (DNA-DNA hybridization) is exploited in the molecular cytogenetic technique, FISH. FISH utilizes fluorescently labeled DNA probes to detect or confirm gene or chromosome abnormalities that are generally beyond the resolution of routine cytogenetics. It is possible to visualize by hybridization, the site of a fragment of DNA of as little as 1 - 2 kb (but more usually 40 - 50 kb) at an efficiency approaching 100%. The sample DNA (metaphase chromosomes or interphase nuclei) is first denatured - a process that separates the complimentary strands within the DNA double helix structure. The fluorescently labeled probe of interest is then added to the denatured sample mixture to hybridize at the target site with the sample DNA as it reanneals back into a double helix. The probe molecule is labeled with a hapten such as biotin. The biotin is located with streptavidin. The streptavidin is located with antibodies. A fluorescent dye may be conjugated to the streptavidin or to the antibodies (probe DNA-biotin-streptavidin-antibody-fluorescent day). When the spread chromosomes are illuminated through a fluorescent microscope, the sample DNA is scored for the presence or absence of the fluorescence signal where the probe/ streptavidin/antibody/fluorescent dye multilayer sandwich has built up. From three to five layers of fluorescent antibodies are built up to amplify the signal. Complex probes made from entire chromosomes can be made (chromosome painting). In this way the presence of a chromosome can be ascertained and whether it has been subject to any rearrangement.

The number of *microdeletion syndromes* diagnosed by FISH is expanding rapidly particularly using the metaphase chromosomal preparations. Although the sensitivity of FISH is better than routine cytogenetics, sensitivity depends on the particular syndrome. In some syndromes, the probe is specific for the defective gene as in Williams Syndrome where a deletion has been shown in the elastin gene in 96% of individuals with a firm diagnosis. In other syndromes such as Prader-Willi/Angelman, the etiology of the syndrome is heterogeneous and microdeletions compose only 60% of the cases. Examples of microdeletion syndromes include:

- Cri-du-Chat
- Miller-Dieker Syndrome
- Smith-Magenis Syndrome
- Steroid Sulfatase Deficiency
- DiGeorge/Velo-Cardio-Facial/CATCH-22/Shprintzen Syndrome
- Kallman Syndrome
- Williams Syndrome
- Wolf-Hirschhorn
- Prader-Willi/Angelman Syndrome.

FISH can be used in interphase cells to determine the chromosome number of one or more chromosomes as well as to detect some specific chromosome rearrangements that are characteristic for certain cancers. The primary advantage of interphase FISH is that it can be performed very rapidly if necessary, usually within 24 hours, because cell growth is not required. A good example is the *Aneuploidy screening test* using amniotic fluid cells in cases with a strong clinical indication for one of the common trisomies. The sample nuclei are denatured and hybridized with DNA probes for chromosomes 13, 18, 21, X, and Y. however, routine cytogenetics is included with the aneuploid screening to confirm the results or detect any abnormalities not detected by interphase FISH (see Figure 31 for an example).

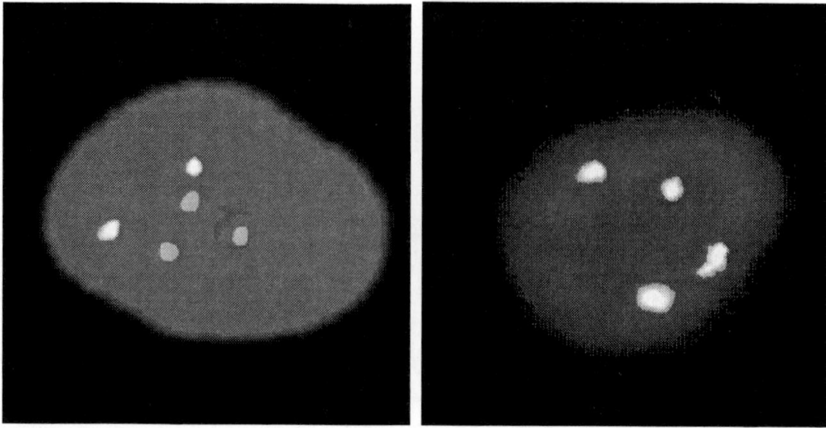

Figure 31. Aneuploidy screening test for a female fetus with trisomy 21 using interphase nuclei from amniotic fluid cells. The nucleus on the left was hybridized to fluorescent DNA probes for chromosomes 13 (green; two signals), and 21 (red; three signals). The nucleus on the right was hybridized to fluorescent DNA probes for chromosomes 18 (aqua; two signals), X (green; two signals), and Y (red; no signal). Therefore, this sample is for a 21 trisomic female.

# The Effect of DNA Mutation

A mutation has a direct effect on the functioning of a genome (gene sequence and expression, and protein product) and an indirect effect on the phenotype of the organism. A mutation that reduces or abolishes a normal protein activity is called *loss-of-function mutation*. They are common but are mostly recessive. However, one intact allele product may not compensate and a disease may ensue, i.e., *haploinsufficiency*, e.g., fibrillin mutation in Marfan syndrome. Moreover, the mutant allele product could paralyze the function of the normal allele product, e.g., the mutant p53 protein. A mutation that confers a new abnormal activity on a protein due to abnormal spatial, temporal, structure-function relationship and/or level of its gene expression is called *gain-of-function mutation*. They are less common but usually dominant, e.g., point mutations affecting the oncogenic c-ras G-protein and c-erbB2.

## Types and Fates of Point Mutations

This type of mutation was chosen for elaboration because is the simplest type of the aforementioned mutations to get the real feeling of how complicated could be the fate or more drastic mutations. Change/alteration of a single base in DNA is known as point mutation as the simplest type of mutation. It affects the regulatory sequences of a gene or its gene proper and hence the mRNA - with several types of consequences. These mutations are either of three types:

- *Homologous, i.e., transitions:* Where a purine base is changed into another purine base, e.g., adenine into guanine or a pyrimidine base into another pyrimidine base, e.g., thymine into cytosine.

- *Heterologous, i.e., transversion:* A purine base is changed into a pyrimidine base and vice versa.
- *Frame shift mutation* with a reading frame shift in the trinucleotide code. It is due to deletion or insertion of a number of nucleotides not divisible by three. Same reading frame shifting occurs during translation and produces an abnormal protein. Insertion or deletion of a number of nucleotides divisible by three produces a protein with an extra or lacking an amino acid(s). This type of mutation has a moderate effect on the produced protein. However, it could have a drastic effect if it targets the active site amino acids, e.g., the most common mutation in cystic fibrosis chloride channel is the deletion of the amino acid number 508 (deltaF508, a phenylalanine).

*Point mutations have four fates (or effects) as follows:*

i. *Silent or synonymous mutation*, in which the new codon specifies the same amino acid as the original codon due to degeneracy of the genetic code. The change lies in the 3$^{rd}$ base of the codon, which differs among codon synonyms for the same amino acid. It will not cause a functional change in the genome. Other forms of silent mutations also include those mutations without an effect on the functioning of the genome because they involve: heterochromatin, intergenic DNA, and, the non-coding components of genes and gene-related sequences without an effect on the gene expression and/or splicing.
ii. *Missense or non-synonymous mutation*, in which the change occurs either in 1$^{st}$ or 2$^{nd}$ base of the codon and produces a different amino acid than the original. The effect of missense mutation on the protein function depends on the position and nature of the replacement amino acid. This could be tolerated; however, a change at the active site will have a greater impact. The protein could be normally functioning (i.e., an *acceptable miss-sense mutation*) such as hemoglobin Milwaukee, which has glutamic acid at position 67 of β-chain instead of the normal valine and hemoglobin Bristol, which has instead an aspartic acid.

    Or, the mutation could be *partially acceptable* because the product will be partially functioning, e.g., *Sickle cell anemia.* Sickle cell anemia or hemoglobin S is an autosomal recessive disease with two normal α-chains but two abnormal β-chains, where an abnormal valine replaces the normal glutamate at position 6. It is partially accepted because HbS binds and releases $O_2$, although inefficiently and become mis-folded, sticky, polymerizes and precipitate as long fibers when it is deoxygenated and at low $O_2$ concentrations. The red cells with HbS are deformed and have a tendency for rapid hemolysis when the mutant gene is homozygous. Presence of normal HbA in the heterozygous state of the disease works as polymerization chain terminator. Patient will appear normal till exposure to inducers, e.g., acidosis, infections, high altitude low $O_2$ tension, hypoxia and dehydration; all increase HbS polymerization. The disease is characterized by hyperbilirubinemia, blood vessel occlusion and clotting. Molecular approaches to treat the disease include; induction of fetal Hb expression, stem cell transplantation and gene therapy.

    Or, the missense mutation could be *unacceptable* because it produces a non-functioning protein or leads to complete loss of the protein because of defective RNA processing. Example is *methemoglobinemia* (HbM or Hb Boston) that has tyrosine replacing histidine 58 (F8) in the α-chains with inability to bind the heam-$Fe^{3+}$ and $O_2$. *Methemoglobinemia* is caused also by hereditary deficiency of the NADH-dependent methemoglobin reductase that reduces metHb-$Fe^{3+}$ into $Fe^{2+}$ and by Hb oxidizing agents such as sulfonamides.

iii. *Non-sense mutation*, in which the altered base change the whole codon meaning into a premature 'non-sense' or termination codon. This leads to pre-mature stopping of protein synthesis and releases a truncated protein or no protein is produced at all. The effect of this on protein activity depends on how much of the polypeptide is lost at what domain. Usually the effect is drastic and the protein is non-functional or mal-functioning. Example of such truncated proteins is the growth factor receptor c-erbB2 oncogene.

iv. *Read-through mutation* converts a stop codon into an amino acid specifying codon that extends the produced polypeptide with additional number of amino acids at its C-terminus. Most proteins can tolerate short extensions without an effect on function, but longer extensions might interfere with folding of the protein and so result in reduced activity.

v. *Frame shift mutation* is reported in the Hb genes causing β-thalassemia and the apoptotic BAX gene in ovarian cancer. Shifting results in:
   - Intron-exon splice junction mutations, e.g., β-globin gene leading to some types of β-thalassemia.
   - Truncated protein if a non-sense codon is developed due to the shift.
   - Garbled translation into totally different protein downstream the shift.
   - A protein may not be produced at all because hnRNA is unable to be processed and is degraded.

## *Different Types of Mutations Cause β-thalassemia*

One disease, e.g., β-thalassemia, may show several kinds of point mutation that could deregulate, e.g., the promoter function; cap site; intron-exon splice junctions; and RNA cleavage/polyadenylation site, other than the frame shifting.

*Worth Noting: Suppressor Mutation*

It happens that a mutation converts an amino acid-specifying codon to a translation termination (stop) codon. However, subsequently another mutation happens at the anticodon of the tRNA for that amino acid that enabled the mutant tRNA to bind the generated stop codon as if it is an amino acid-specifying codon. This is why it was called a suppressor mutation; i.e., it suppresses the effect of the first mutation. Suppressor mutation is also a subsequent mutation in one gene that reverses the effect of a mutation in the same or another gene. A mutation in the bacterium could make it unable to synthesize a specific metabolite. However, after exposure to a mutagen, the bacterium may acquire another mutation that reverses the effect of the first mutation to regain the original phenotype characteristics. See also the Ames test mentioned above.

# Mechanisms of Repair of DNA Damage

It is worth noting that ~¼ of our genes is used to replicate, express and maintain our genome. Cellular DNA is critical for cellular control and replication and necessitates continuous surveillance and repair utilizing efficient plethora of damage response mechanisms to maintain genomic stability. These include processes that recognize and repair the damage, cell cycle checkpoints that prevent cell cycle progression in the presence of damage, and mechanisms, such as apoptosis as a final fate for a damaged cell with irreparable DNA.

Three major repair mechanisms exist: the direct reversal repair; the excision repair that includes the base and nucleotide excision repairs and mismatch repair; and the double strand breaks repair.

1. *Direct Reversal Repair:*

   Very few lesions are directly repaired by simple reversal without nucleotide replacement and include:

   - *Simple nick* without any complicating damage to surroundings as those caused by the direct effect of ionizing radiation are simply resealed by *DNA ligase*.
   - Pyrimidine dimers (particularly thymine-thymine dimer) occur spontaneously, or induced by ultraviolet (UV) light, chemicals (e.g., tobacco smoking byproducts), or radiation. *Pyrimidine dimers* are repaired by the light-dependent direct photoreactivation involving *DNA photolyase* that is stimulated by 300 - 500 nm light in *E. Coli* and lower eukaryotes.
   - *Methylation of guanine* by alkylating agents is repaired by the "suicidal" *$O^6$–methylguanine-DNA methyltransferase* that transfers the methyl group from $O^6$–methylguanine into its own cysteine residue. This leads to irreversible inactivation of the enzyme. The gene of this enzyme is frequently repressed by hypermethylation in colon cancer. This permits the alkylating agents-induced GC-AT conversions. The later is implicated in the activation of the proto-oncogene K-ras into its oncogenic form seen in about half of colorectal carcinomas. This enzyme removes also chloroethyl and benzyl groups from guanine.

2. *Excision repair:*

   Excision repair is the most active DNA repair system that utilizes three mechanisms; base excision repair, nucleotide excision repair, and, mismatch repair. They act to correct many types of DNA damage affecting only one strand by replacing the damaged site by a new synthesis using the other strand as template.

   A. *Base excision repair for single nucleotide damage:* It repairs minor damaged nucleotides induced spontaneously, or by base analogs incorporation, alkylation and/or by oxidation due to ionizing radiation in the mitochondria and nucleus. The single nucleotide damages also include: hydrolytic (spontaneous) or oxidative deamination products (uracil from deaminated cytosine; xanthine from deaminated guanine and hypoxanthine from deaminated adenine), oxidation products (e.g., 8-oxo-deoxyguanine, 5-hydroxycytosine and thymine glycol), and methylation products (e.g., 3-methyladenine, 7-methylguanine and 2-

methylcytosine). Spontaneous depurination occur at 37 °C at rate of 5000-10,000/cell/day due to the thermal liability of the purine N-glycosidic bond.

These defects are repaired by initial cleavage of the β-*N*-glycosidic bond between the damaged base and the sugar catalyzed by one of the *specific-DNA N-glycosylases* leaving apyrimidinic or apurinic (AP) site. Thus, the system also repairs apyrimidinic or apurinic (AP) site. The limited substrate specificity of each DNA glycosylases possessed by a cell limits the range of damaged nucleotides that can be repaired by the base excision repair.

An *AP-endonuclease* cleaves the 5'-phosphodiester bond of the target sugar and another phosphodiesterase cuts its 3'-bond. Alternatively, the 3'-bond cleavage could be done by the endonuclease activity of the N-glycosylase while removing the base followed by the phosphodiesterase 5'-cut to remove the sugar residues. Then, *DNA polymerase I (β* in mammals or *δ* in yeast*)* inserts the proper nucleotide, dictated by the base on the undamaged complementary strand. Then, the deoxyribose-phosphate backbone is rejoined by a *Ligase*; see, Figure 32. Since this system introduces single strand break, it is also important in repair of the single strand break and is deleterious if became over-functional, i.e., it may introduce single strand breaks. Irreparable AP sites decay to single-strand breaks.

*Worth Noting: Two extremes cause no inherited defects*

There are no known diseases associated with inherited defects in the base excision repair. This may be due to the many redundant glycosylases and/or the life-incompatibility of mutations affecting the other proteins that lead to early embryonic death.

The above mentioned types of single-strand DNA damage highly induces the activity of the *NAD-dependent Poly(ADP-Ribose) Polymerase-1* (PARP-1). PARP-1 covalently modifies histones and the DNA nuclear regulatory proteins by adding many ADP-ribose molecules from $NAD^+$ ($NAD^+$ is converted into nicotinamide) to their glutamic acid residues into long and branched polymers. This cause chromatin remodeling through increasing the negative charge of these proteins to release the bound DNA and de-condensing chromatin that becomes accessible for DNA repair enzymes and transcription factors.

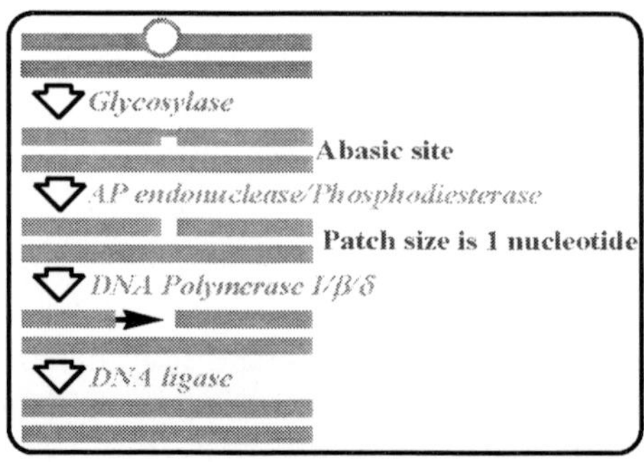


Figure 32. The mechanism of the base excision repair with excised patch of one nucleotide size.

*Worth Noting: NAD-dependent Poly(ADP-Ribose) Polymerase-1 (PARP-1):*

PARP-1 in itself is susceptible to activating ADP-ribosylation by other PARPs. Poly(ADP-ribosylation) is a reversible covalent protein modification that is rapidly reversed by poly(ADP-ribose) glycohyrolase. Alternative mechanism of PARP-1 activation by phosphorylated externally regulated kinase (ERK) implicates PARP-1 in a vast number of signal-transduction networks in the cell. Therefore, PARP inhibitors prevent PARP-1 activation in response to nicked DNA, in an attempt to suppress PARP-mediated DNA repair. Although they differ in their chemical structure, their potency, stability, solubility in water, and therapeutic potential, these inhibitors block the binding of the nicotinamide moiety of $NAD^+$ in the catalytic site of the enzyme. Several groups of PARP inhibitors (including phenanthridine derivatives) were designed to protect cells under stress conditions from cell death induced by a massive activation of PARP-1, e.g., during stroke and inflammation. Oppositely, because polyADP-ribosylation is required for spindle formation these PARP inhibitors cause G2/M cell cycle arrest raising the potential as anticancer drugs. Cancer cells are particularly sensitive to this effect whereas, normal cells are resistant.

*B. Nucleotide excision repair:* Although it is a backup system for the base excision repair, it differs from base excision repair in that: it is only nuclear, having broader specificity; having the ability to deal with more extreme forms of damage such as intra-strand cross-links (e.g., pyrimidine dimers pyrimidine dimers and purines-deoxyribose-phosphate cross links due to the hydroxyl radical) and bulky base adducts (e.g., those caused by aflatoxins); does not require initial affect base removal, and a longer stretch of DNA including the damage is excised.

This mechanism replaces the photoreactivation system for repair of the pyrimidine dimers in organisms devoid of photoreactivation system, e.g., humans. It also repairs strand breaks, inter-strand base cross-links or DNA-proteins cross-links and base modifications (e.g., the tobacco smoking-induced benzo[a]pyrene-guanine adducts). It repairs DNA strand damage of 2 – 30 bases in length. The pyrimidine dimers, bulky base adducts and cross-links prevent progress of the RNA and DNA polymerases beyond the lesion site.

*Nucleotide excision repair is of two subtypes:*

a. Global-Genome Repair.
b. Transcription-Coupled Repair.

The *slower Global-Genome Repair* works on the whole exposed genome by recognizing strand defects with XP proteins (their mutation leads to *Xeroderma Pigmentosum*). The XP helicase activity unwinds the damaged DNA sequence to be accessible to the other repair enzymes.

*Transcription-Coupled DNA Repair* preferentially repairs genes and particularly those that are actively transcribed as compared to other genome sequences. A component of this repair is directly linked to the basal transcription process, and is termed "strand specific" DNA repair, since the transcribed strand of the active genes is preferentially repaired. Transcription-coupled repair is induced by detecting a stalled RNA polymerase - which is unable to proceed because of the DNA strand damage in the target gene. The detection proteins are

CS proteins (A and B; their mutation leads to the aging disorder, *Cockayne Syndrome*) and p53. CS proteins displace the stalled RNA polymerase and XP helicase unwinds the DNA to allow an access to excision repair enzymes.

In *E. Coli* short patch repair, the three subunits of the *eximuclease (UV-specific endonuclease; a special excision nuclease)* recognize distorted DNA due to the damage and cut the sequence on either side of the damage ($5^{th}$-nucleotide downstream and $8^{th}$-upstream). The 12-oligonucleotide containing the lesion is displaced and diffuses intact by a helicase action. In eukaryotes the process is similar but is called long patch because it excises 24-29 nucleotide sequence utilizing ~20 proteins ($5^{th}$-nucleotide downstream cut and a further upstream cut). Cutting uses endonucleases that cut single-stranded DNA at its junction with a double-stranded DNA, as is the case at the end of a transcription bubble. *DNA polymerase I (δ/ε in human)* fills the gap with new nucleotides using the healthy strand as a template in the 5'⇒3' direction. *DNA ligase* joins the 3'-end of the new DNA and the 5'-end of the original DNA; see Figure 33.

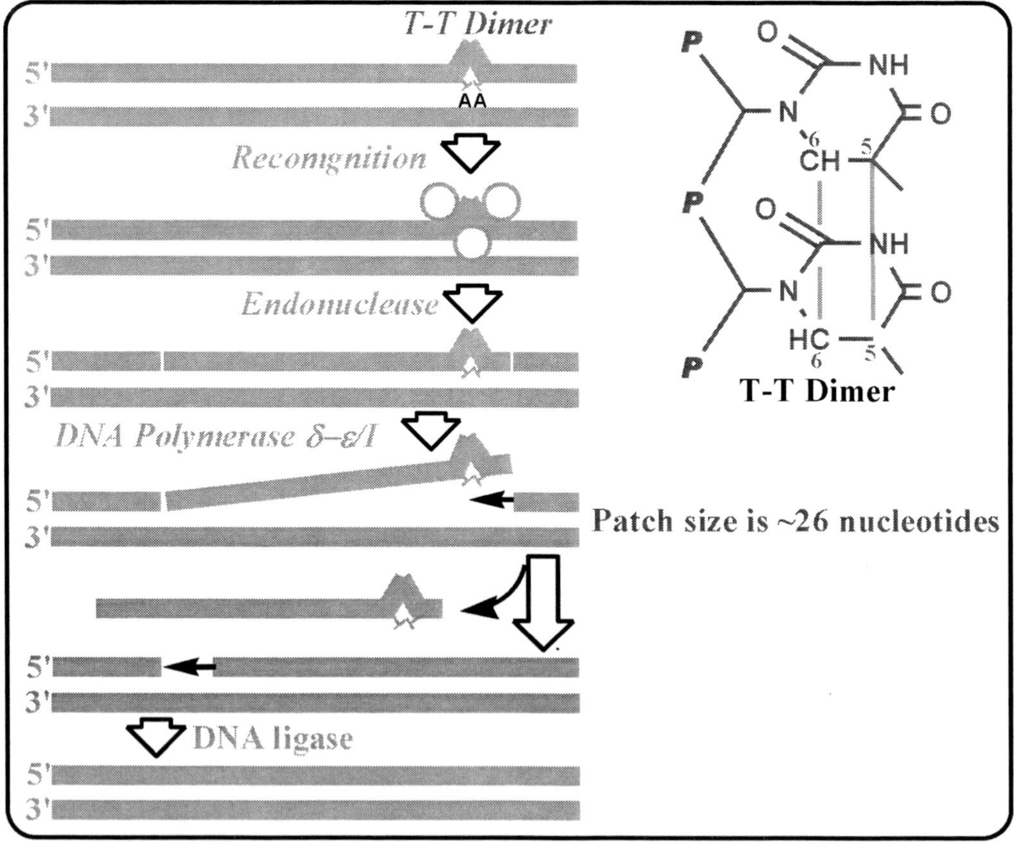


Figure 33. The mechanism of nucleotide excision repair. The insert shows the structure of the thymine-thymine dimer.

*Worth Noting: Nucleotide excision repair is error-prone*

The tumor-suppressor protein *BRCA1* (breast cancer susceptibility gene 1; often mutant in *breast and ovarian cancers*) is essential for Transcription-Coupled Repair associated with oxidative DNA damage but not ultraviolet light damage. Cells from Cockayne and Werner's aging syndromes are deficient in the transcription-coupled DNA repair. The many steps and more than 20 proteins involved in unwinding the DNA and recognizing the damage make this system error-prone than the base excision repair. A defect in this system was ascribed to *hereditary non-polyposis colorectal cancer*.

*C. Mismatch Repair:* Mismatch, e.g., G-T pairing, is due to copying errors during replication, which can produce frame shift mutations. This system can correct A-C and T-C mismatches more efficiently than G-A and T-C mismatches, and is very poor correcting C-C mismatches. Mismatch repair plays a significant role in protecting against incorporation of 8−hydroxy/8−oxodeoxyguanosine into DNA. Functional mismatch repair also prevents the alternative telomerase-independent lengthening of telomeres and so protects against cancer.

In bacteria long patch system, the mismatch is recognized by a group of proteins called *MutS*, *MutL* and *MutH* (with a 5'-GATC-3' endonuclease, plus 5 more proteins in eukaryotes) that scan the newly synthesized daughter DNA strand for improper base-pairing. The system identifies the parent DNA strand (old) by its 6-methyladenine in the 5'-GATC-3' sequences, whereas, the new strand is not methylated yet. Once they have located the mismatch or unpaired DNA sequence (e.g., loops of 1 to 5 of unpaired bases due to slippage of DNA polymerase) in daughter unmethylated strand, they nick it on either 3'- or 5'-side of the defect at a site upstream the G in the corresponding 5'-GATC-3' sequences of the old strand flanking the defect. This is done with the help of a helicase activity. The patch of DNA cleared is ~1 kb. Other short and very short repair systems utilize alternative proteins and clear a patch of ≤10 nucleotides.

A special exonuclease (*exonuclease I*) hydrolyzes DNA into free DNA nucleotides from the nick in 3'⇒5' or 5'⇒3' direction to a few nucleotides after the defect. A new DNA is synthesized to fill the gap by *DNA polymerase III* and *DNA ligase* joins the 3'-end of the new DNA and the 5'-end of the DNA ahead of the defect; see Figure 34.

*Worth Noting: Human homologues of MutS and MutL proteins*

Eukaryotic system works the same and deficiency in human homologues of MutS and MutL proteins (hMSH2 and hMLH1, respectively) lead to a specific hereditary type of cancer, i.e., *hereditary non-polyposis colorectal cancer*. In this disease, the defective repair leads to failure of removal of unpaired DNA sequences and hence microsatellite expansion and instability. This culminates into defective cell cycle control and cancer.

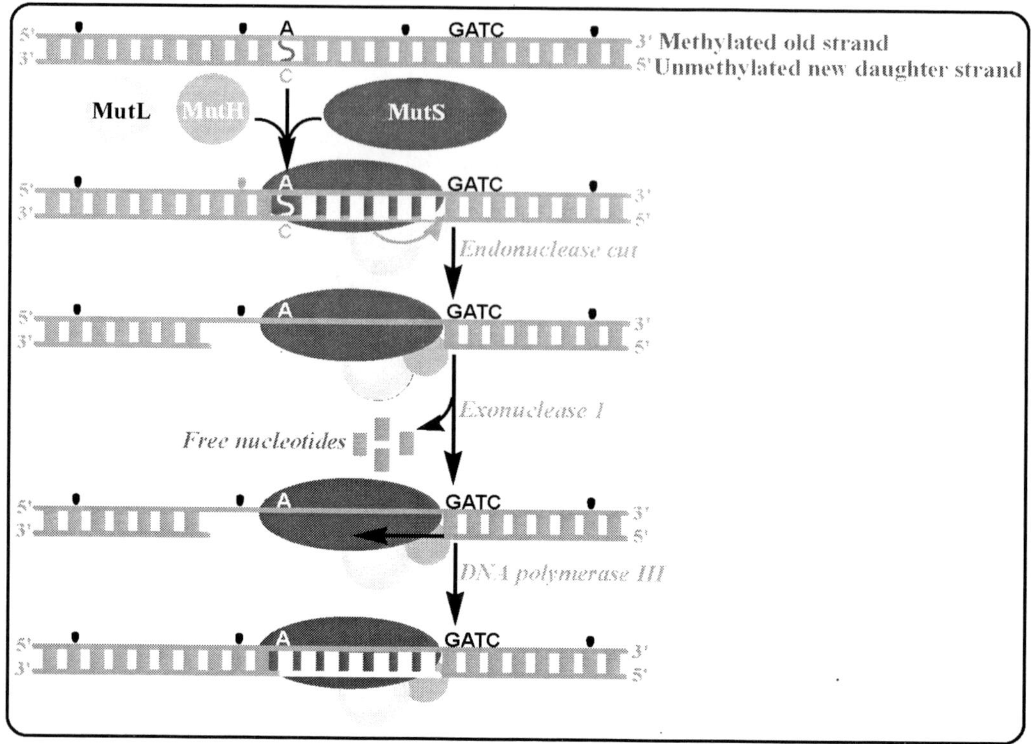


Figure 34. The mechanism of the mismatch repair.

3. *Double strand breaks repair:*

Single strand breaks are not critical to the cell and are repaired by PARP-1-activated template-dependent DNA synthesis. The more serious double strand beaks are due to ionizing radiation, strand breaking and alkylating chemotherapy and oxidative free radicals. Double strand breaks are also normally created in a controlled fashion by the cell during recombination events, e.g., the genome rearrangements of the immunoglobulin gene segments. Although double-strand breaks are rare, they are difficult to repair and can be very injurious for dividing somatic cells. Repair of double-strand damage in mammals can best occur during meiotic cell division of germ cells when homologous pairs of chromosomes line-up, while; complementary pairs don't typically line-up very well in mitotic somatic cells.

*Double strand beaks are repaired by two mechanisms:*

- *Non-Homologous End-Joining*, the simplest but is the least accurate due to rejoining of the unrelated two broken ends without regard to deletions, mismatching or rearrangements.
- *Homologous Recombination*, which is less frequently used but more accurate by exactly reconstituting the original sequence using the homologous chromosome or the sister chromatid as template: see Figure 35.

In the *non-homologous end-joining*, two proteins are involved; the heterodimer *Ku* (Ku70/Ku86; an ATP-dependent helicase) and *DNA-dependent protein kinase*-CS. The *WRN protein* (it is mutated in the aging disorder, *Werner's syndrome*), operates in both types of double strand breaks repair. The Ku protein initiates non-homologous end-joining by binding to each of the broken DNA ends and bringing them together. DNA-dependent Protein Kinase and WRN are recruited to bind Ku and the broken ends. The DNA-dependent Protein Kinase approximates the two broken ends by an internal DNA binding site. The complex activates the kinase that in turn activates - by phosphorylation - the Ku; and all recruit the DNA ligase.

The activated Ku/WRN unwinds the double stranded helix on both ends. This allows base pairing between two opposite stands from both ends leaving the other two tails free. The extranucleotide tails are degraded into free nucleotides by exonuclease and the gaps are filled by DNA polymerase III and DNA ligase seals the free ends. Ku86 and DNA-dependent Protein Kinase are also important for telomere maintenance. The rearrangement of the immunoglobulin gene is carried out by a similar mechanism. Although non-homologous end-joining is very error-prone, it is often effective because the joining junction mostly involve non-coding sequences. *Non-homologous end-joining typically occurs in the G1 phase of the cell-cycle when the relevant chromosome is one chromatid.* The tumor-suppressor protein *BRCA1* has also been shown to play a critical role in non-homologous end-joining. Protein kinases, e.g., the Ataxia Telangiectasia (ATM) and p53 proteins are other proteins involved to stop the cell cycle until the repair takes place.

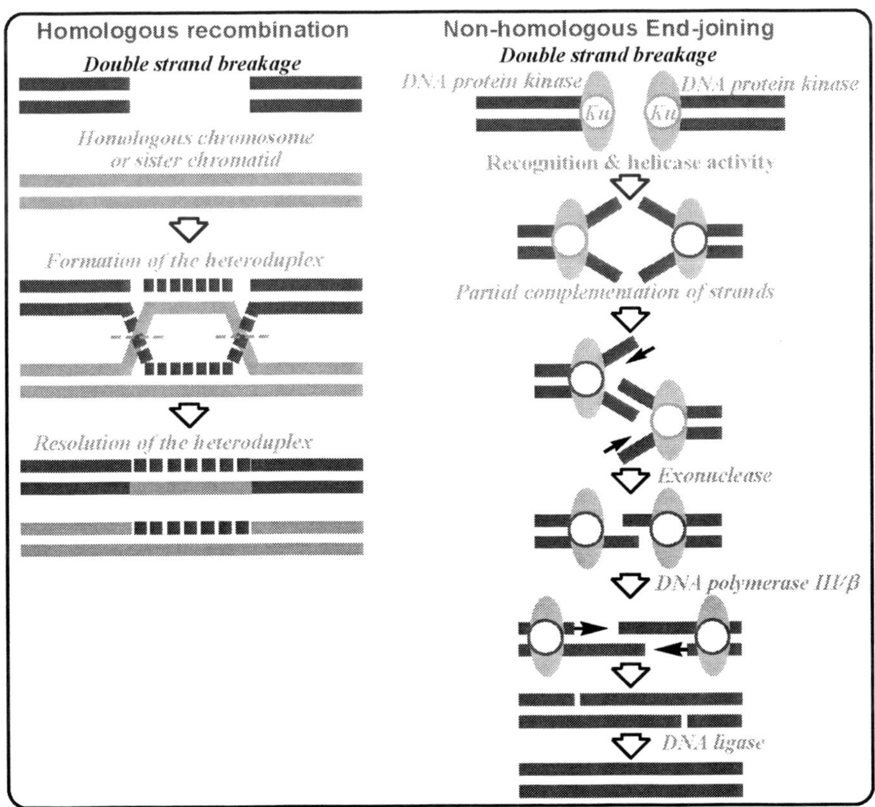


Figure 35. The two repair types of DNA double strand break. On the left is the homologous recombination utilizing sister chromatid strands as template. On the right is the non-homologous end joining with a dominant role for Ku and the DNA-dependent protein kinase.

*The homologous recombination* is the dominant method used in the S and G2 phases of the cell-cycle, after a sister chromatid has been created. Homologous pairing of the sister chromatids is often mediated by *Rad51 protein*, which is also necessary for cell proliferation and survival. *Rad52 protein* recognizes the double strand breaks and adheres to the free ends of the break while Rad51 searches the undamaged sister chromatid for a homologous repair template. The tumor suppressor protein *BRCA2* co-localizes with Rad51 during homologous re-combinational repair, and contributes significantly to its activity. Homologous recombination could be the base for telomerase-independent telomere lengthening in mammals. The presence of multiple genomes for mitochondrial DNA makes homologous recombination more feasible for repairing its double strand breaks.

*From the mutation repair discussion one may notice the integration of it with cell cycle control, transcription and gene rearrangement through common proteins.*

## Diseases Related to Defective Repair of DNA Damage

*Xeroderma Pigmentosum (XP):* It is an autosomal recessive disease due to *defective nucleotide excision repair* caused by mutation of at least 7 genes called *XPA – XPG* with helicase, kinase, recognition and excision activities. They are also subunits in the basic transcription machinery. Patients are highly sensitive to sunlight and UV lights and their skin is damaged and is liable for multiple cancer, immunodeficiency, atrophy and scarring of eyelids, ulcerated cornea, and/or progressive neurological degeneration and may lead to premature death.

*Ataxia Telangiectasia (AT):* It is caused by autosomal recessive mutations in the ATM gene (for Ataxia–Telangiectasia Mutated). ATM is a protein kinase known to be an ''early responder'' to double-stranded DNA breaks and its deficiency leads to increased sensitivity to X and other ionizing radiation-induced damage. In presence of such DNA damage, normal ATM protein (mainly acting through p53 protein) blocks progression through the cell cycle to induce repair or apoptosis. Cells of AT patients are resistant to apoptosis and do not undergo cell cycle arrest when subjected to ionizing radiation. Cerebellar degeneration, immunodeficiency, sterility, increased risk of cancer of the reticuloendothelial system and genomic instability are the characteristics of the disease. Patients have a lifespan of only about 20 years.

*Hereditary Non-polyposis Colorectal Cancer (HPCC) and Other Types of Cancer:* It is also called Lynch syndrome in which patients develop cancer in the colon along with cancer of stomach, ovary and/or uterus. It is due to defective DNA mismatch repair with mutations in two genes, hMSH2 and hMLH1 (the human analogs of MutS and L). It has an autosomal dominant pattern of inheritance with cancer at early age. Another mutation ascribed to this disease involves defective transcription-coupled nucleotide excision repair.

*Cockayne syndrome (CS):* Cockayne syndrome (after, Edward Alfred Cockayne; first patient to be characterized) is a rare inherited disorder in which people are sensitive to sunlight, have short stature, and have the appearance of premature aging. In the classical form of Cockayne syndrome (Type I), the symptoms are progressive and typically become apparent after the age of 1 year. An early onset or the congenital form of Cockayne syndrome (Type II) is apparent at birth. Interestingly, unlike other DNA repair diseases, Cockayne syndrome is not linked to cancer. There is *defective transcription-coupled nucleotide excision repair* after exposure to UV radiation. Two genes defective in Cockayne syndrome, *CSA* and *CSB*, have been identified so far. Both genes code for proteins that interacts with components of the transcriptional machinery and with DNA repair proteins.

*Werner syndrome (WRN or Progeria of the adult):* It is an autosomal recessive syndrome caused by null mutations at a locus (WRN) coding for a member of the RecQ family of helicases. The gene product could play a role in such processes as DNA replication, DNA repair, DNA recombination, gene transcription and chromosome segregation. The presence of a 3'-5' exonuclease domain in the protein is consistent with a role in DNA repair. While WS is extremely rare (1-22 per million), the prevalence of heterozygous individuals may be as high as 1/150 to 1/200 individuals, at least in Japan. Somatic cells from such individuals have been shown to be hypersensitive to a powerful carcinogen and genotoxin, 4-nitroquinoline-1-oxide. The family members of WS pedigrees may be at higher risks for malignancy.

*Fanconi's anemia:* It is an autosomal recessive anemia associated with increased susceptibility to chromosomal instability, pancytopenia and acute myeloid leukemia. It is due to defective correction of *chemical-induced DNA cross-links*. The FANC (A, B, C, D1, D2, E, F, G, I, J, and L - Fanconi's complementation groups) tumor suppressor genes are located to 11 complementation groups (e.g., FANCA at 16q24.3; FANCC at 9q22.3; FANCD2 at 3p25.3; FANCE at 11p15). They include components of DNA repair machinery where FANCD1 is the same for BRCA2.

*Fragile X syndrome:* Fragile X Syndrome is probably the commonest single-gene cause of learning disability in humans with an estimated prevalence of 1/250 males, where it causes moderate to severe intellectual and social impairment together with large ears and head, long face and macroorchidism. A fragile site *(FRAXA)* is expressible at the gene locus at Xq27.3 with a large expansion (in the full mutation) in a 5'-CGG-3' microsatellite repeat tract in the first un-translated exon of the *FMR1* gene. This tract is normally composed of 30 copies but abnormally becomes thousands of copies.

This gene normally encodes the RNA-binding protein FMRP. The mutation disrupts the gene and induces its hypermethylation. Smaller expansions (permutations) do not cause Fragile X syndrome but may progress into full mutations over one or more generations. The disease has variable phenotype in females (from asymptomatic to full disease – 1/2000) possibly due to differences in the proportions of active and inactive normal and mutated X chromosomes in the relevant tissues. A small minority of Fragile X cases is due to point mutations or deletions in the coding sequence rather than CGG repeat expansions and there is no fragile site expression or hypermethylation.

The disease is called so because X chromosome develops DNA single and double-stranded breaks in cases with folic acid deficiency. This vitamin is essential for synthesis of nucleotides and DNA replication and repair.

## Mitochondrial DNA

Most animal cells contain between a few hundred and a few thousand mitochondria. The most mitochondria are found in the cells that are most metabolically active: neurons and muscle cells, where mitochondria make up about 40% of cell volume. Hence mitochondrial diseases are often encephalomyopathies, e.g., Mitochondria Encephalomyopathy, Lactic Acidosis, and Stroke. About 10% of the body weight of a human adult is mitochondria that generate 95% of the body energy. The human mitochondrial genome is less than 0.1% of the total cell DNA. It was completely sequenced and genes were identified. There are 2 - 10 copies of a supercoiled, double-stranded circular molecules of mitochondrial DNA (mtDNA); each of 16,569 bps in size containing only 37 genes. 13 genes code for 13 subunits of the 87 proteins required for oxidative phosphorylation. They include 6 subunits of NADH-dehydrogenase complexes I; 1 subunit for cytochrome b, 3 subunits of cytochrome oxidase complex IV and 2 subunits of ATP synthase $F_0$ complex V. 24 genes code for the non-coding RNAs: 12S and 16S rRNAs and 22 tRNAs required for protein synthesis in the mitochondria. Therefore, most of ~1,000 proteins required for oxidative phosphorylation and mitochondrial translocators are encoded by nuclear DNA (nDNA). Moreover, the proteins required for; coordination of the expression of nuclear and mtDNA, importation of proteins into the mitochondria, assembly of the mitochondrial complexes, and, regulation of mitochondrial fission are encoded by nuclear genes.

The genes in this genome are much more closely packed than in the nuclear genome (see Figure 36). They do not contain introns and some are overlapping (e.g., ATP6 and ATP8 genes for ATP synthase subunits 6 and 8). Within a cell, mtDNA replication is semi-autonomous and occurs mainly during G1 phase of cell cycle and is not synchronized with the S phase of the cell cycle. However, the two genetic systems (nuclear and mitochondrial) appear to be closely coordinated by yet unknown mechanisms. The overall biogenesis of mitochondria keeps pace with cell proliferation. The mitochondria divide by fission during G1 phase of cell cycle.

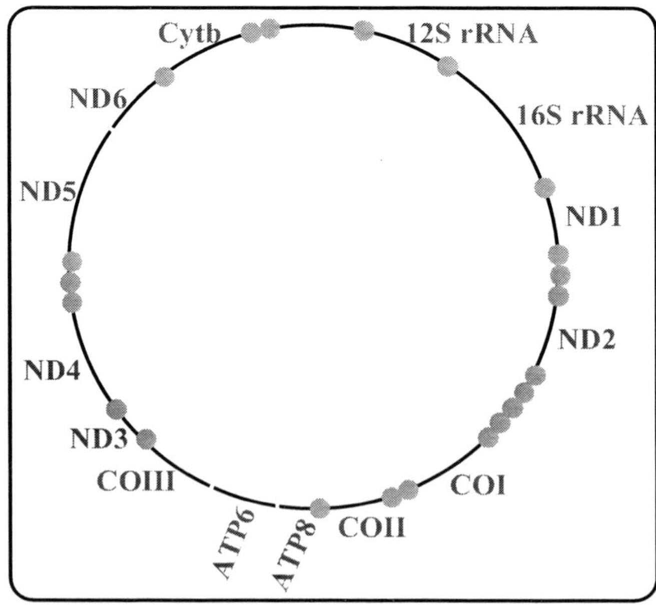


Figure 36. The map of the human mitochondrial DNA. ATP6 and ATP8 are the genes for ATPase subunits 6 and 8. COI, COII and COIII are the genes for cytochrome c oxidase subunits I, II and III. Cytb is the gene for apocytochrome b. ND1 - ND6 are the genes for NADH-hydrogenase subunits 1- 6. Ribosomal RNA (12S rRNA and 16S rRNA) and transfer RNA (•) are two types of non-coding RNA.

The translation of the mtDNA uses a slightly different genetic code than the nDNA translation (the stop UGA is read tryptophan and the arginine; AGA and AGG are read stop). Mitochondrial RNA transcription is preferentially inhibited by ethidium, whereas, transcription of RNA from nDNA is preferentially inhibited by α-amanitin. Likewise, protein synthesis in mitochondria is sensitive to inhibition by aminoglycosides, tetracycline and chloramphenicol, whereas, cytosolic protein synthesis is selectively inhibited by cycloheximide. This makes the mitochondrial transcription and translation systems resembling the prokaryotic systems.

The mitochondrial DNA has high mutation (10-20 times the rate seen in nuclear DNA) rate due to lack of histone protection; imperfect repair and the proximity of the mitochondrial DNA to free radical producing respiratory chain reactions. However, recombination of mitochondrial DNA molecules between fused mitochondria buffer the effect of such mutation but create massive heterogeneity in the mitochondrial genome (called heteroplasmy). Therefore, the biological impact of a given mutation may vary, depending on the proportion of mutant vs. wild type mitochondrial DNAs carried by the individual patients and even among different tissues of the same patient. This effect contributes to the various phenotypes observed amongst family members carrying the same pathogenic mitochondrial DNA mutation. Since most of mitochondria in the zygote come from the ovum (hundreds of thousands) and very few come from the sperm (a few hundred), mutations in mitochondrial genes and related diseases have a maternal pattern of non-Mendelian inheritance. And, analysis of mitochondrial DNA is used for forensic determination of motherhood relatives. Mutations in mitochondrial DNA involve deletions, duplications, or point mutations. Symptoms of these defects would appear in one or more of the tissues with the highest ATP demands: nervous tissue, heart, skeletal muscle, and kidney. They lead to decreased activity of the electron transport chain and energy production and increased accumulation of lactate (lactic acidosis and increased blood lactate/pyruvate ratio) that causes the most commonly encountered degenerative diseases. Mitochondrial dysfunction is also implicated in Parkinson's disease, dilated and hypertrophic cardiomyopathies, diabetes mellitus, Alzheimer disease, depressive disorders, etc.

*Worth Noting: Inherited mitochondrial disorders (OXPHOS)*

Oxphos diseases are genetically inherited diseases involving the components of oxidative phosphorylation (i.e., Oxphos) due to mutations in mtDNA encoding a few mitochondrial proteins, rRNAs and tRNAs, or nDNA encoding most of the mitochondrial proteins.

Though clinically and genetically heterogeneous, mitochondrial disorders are characterised by biochemical abnormalities of the mitochondrial respiratory chain, due to mitochondrial mutations or nuclear genome mutations (e.g., genes *POLG*, *ANT1* or *Twinkle*). The nuclear mutations do not show a pattern of maternal inheritance but are usually autosomal recessive. The mutations are uniformly distributed to daughter cells and therefore are expressed in all tissues containing the allele for a particular tissue-specific isoform. However, phenotypic expression still will be most apparent in tissues with high ATP requirements.

The diseases due to mitochondrial mutations include Chronic Progressive External Ophthalmoplegia (CPEO) and Kearns Sayre Syndrome (KSS) which are typified by external ophthalmoplegia, ptosis and mitochondrial myopathy. They are due to large-scale mtDNA rearrangements involving tRNAs and respiratory chain genes (both single and multiple mtDNA deletions and duplications). This group also include Pearson syndrome, a systemic disorder of oxidative phosphorylation that predominantly affects bone marrow.

Mitochondrial Encephalomyopathy with Lactic Acidosis and Stroke-like episodes (MELAS), Myoclonic Epilepsy with Ragged-Red Fibres (MERRF) and Neurogenic weakness with Ataxia and Retinitis Pigmentosa (NARP) are all associated with common point mutations within the mitochondrial genome (3243A>G MELAS, 8344A>G MERRF and 8993T>G/C NARP) that involve tRNAs ($tRNA_{lys/leu}$) and rRNAs. The MERRF disease is typified by late childhood or adult onset myopathy with ragged-red fibers, and a slowly progressive dementia. The MELAS disease is typified by progressive neurodegenerative disease with stroke-like episodes first occurring between 5 -15 years of age and a mitochondrial myopathy.

Three primary mtDNA missense point mutations involving mitochondrially encode subunits of NADH dehydrogenase complex I (3460G>A *MTND1*, 11778G>A *MTND4* and 14484T>C *MTND6*) cause Leber's Hereditary Optic Neuropathy (LHON). Similar diseases involving mtDNA missense mutations in respiratory chain polypeptides include Leigh disease (sub-acute necrotizing encephalopathy) that mainly affects $F_0$ subunits of ATP synthase. The disease is typified by optic atrophy, ophthalmoplegia, nystagmus, respiratory abnormalities, ataxia, hypotonia, spasticity, and developmental delay or regression.

## Mitochondria and Aging

Like it was with the telomeric theory of aging, there is the mitochondrial theory of aging that is an integral part of the oxidative stress theory. The commensalism treaty between the mitochondrial and the formed eukaryotic cell is functional, there should not be problems. Problem happens when either the nuclear serves to mitochondrial gets suboptimum, as mentioned above for mutations in mitochondria-serving nuclear genes, or when the mitochondrial serves to the cell gets suboptimum through generating less energy, releasing more free radicals, disobeying the regulated mitochondrial recycling through autophagy by the lysosome, and/or initiating in the dismissal fate of the cell through apoptosis. Transgenic mice that have defective repair of mtDNA have reduced lifespan, increased apoptosis and display an "accelerated aging" phenotype.

*An estimated 1-2% of oxygen used by mitochondria will normally "leak" from the respiratory chain to form Superoxide anions ($O_2^{\cdot -}$).* Defects in the oxidative phosphorylation enzyme complexes reduce energy production and increase reactive $O_2$ species production and toxicity. Superoxides inactivate the mitochondrial enzymes containing iron-sulfur centers including: aconitase, succinate dehydrogenase and NADH dehydrogenase. Not only oxygen free radicals but also nitric oxide ($NO^{\cdot}$) is produced - giving the chance for the formation of the very damaging peroxynitrite.

Aging of the mitochondria affect mainly complex I and III because they are mitochondrially-encoded. Complex I (with 7 mtDNA-coded proteins; more than a quarter of all the proteins in the complex) generates $O_2^{\cdot -}$ in one of the iron-sulfur clusters that goes to the mitochondrial matrix where it could be neutralized by the mitochondrial Mn-superoxide dismutase. Most of the $O_2^{\cdot -}$ generated from Complex III comes from $CoQ^{\cdot -}$; half goes to the matrix to be neutralized and half floats toward the cytoplasm. Cytochrome c oxidase activity declines with age, resulting in increased production of superoxide and hydrogen peroxide.

The mitochondrial membrane potential ($\Delta\Psi$; expressed in millivolts - mV) is gradually reduced in the early stage of the programmed cell death by apoptosis and is abnormal in transformed cancer cells. In such a condition, apoptosis is due to the leakage of cytochrome c into the cytoplasm. Cytochrome c is a strong endogenous promoter for apoptosis machinery. Not only the mitochondrial efficiency and its membrane potential are abnormal with early cancer cell transformation, but also the mitochondrial mass is reduced. This acquires cancer cells an anaerobic metabolism with excessive lactate production in aerobic condition. During acute oxygen deprivation, e.g., from ischemia (a low blood flow), the sharp inability to generate ATP from the electron transport chain results in an increased permeability of the inner mitochondrial membrane leading to mitochondrial swelling and swelling of the other cell organelles and the whole cell. Mitochondrial swelling is a key element in the pathogenesis of irreversible cell injury leading to cell lysis and death through necrosis.

The mitochondrial mass per cell differs not only because of the cell type but also depending on the developmental, metabolic and disease status of the cell, e.g., cancer transformation is correlated with reduction in the mitochondrial mass. The mitochondrial mass is measured by several methods including the staining of the ethanol-fixed cells with mitochondria-specific stains, e.g., acridines orange 10-nonyl bromide followed by flowcytometric analysis of the mean green fluorescence relative to cell number at 488 nm excitation and 530 nm emission wavelengths. The pattern of the change in the mitochondrial membrane potential is specific to cell state regarding apoptosis and necrosis. Whereas, there is a sharp loss of the potential in the later, apoptosis is accompanied by gradual loss of the potential. The mitochondrial membrane potential is investigated by several ways including staining of fresh cells with rhodamine 123 followed by flowcytometric analysis of the mean green fluorescence in relation to cell number using 488 nm excitation and 530 nm emission wavelengths.

Cell death by necrosis vs. apoptosis depends on the nature and mechanism of mitochondrial permeability transition pore to release cytochrome c. Release of cytochrome c into the cytoplasm is either $Ca^{2+}$-dependent or -independent. The $Ca^{2+}$-dependent release of cytochrome c during necrosis is triggered by mitochondrial $Ca^{2+}$ overload that permanently mitochondrial permeability transition pore through the two layers of the mitochondrial membrane. Cyclophilin D sensitizes the other pore components to $Ca^{2+}$. The entry of large solutes and water into the matrix causes the mitochondrion to swell and burst, releasing cytochrome c into the cytoplasm. ATP energy is depleted if the mitochondrial permeability transition pore (MPTP) remains open. The $Ca^{2+}$-independent release of cytochrome c during apoptosis, the release is due to temporary pore formation only in the outer membrane induced by interaction of Bcl-2/Bax and the outer membrane voltage-dependant anion channel (VDAC). The apoptosis required ATP is ensured by the intact inner membrane (Figure 37).

Cytochrome c is normally held to the inner mitochondrial membrane by the phospholipid cardiolipin (diphosphatidylglycerol) that composes 10% of the inner mitochondrial membrane. Cardiolipin is present at lower concentrations in the outer mitochondrial membrane and is not found in non-mitochondrial membranes. Mitochondrial membrane cardiolipin content declines with age, resulting in a decline in cytochrome c activity. The age-dependent reduction in cardiolipin correlates similar reduction in cytochrome c activity. Free radical-mediated oxidation of cardiolipin releases cytochrome c from the inner mitochondrial membrane to the intermembrane space.

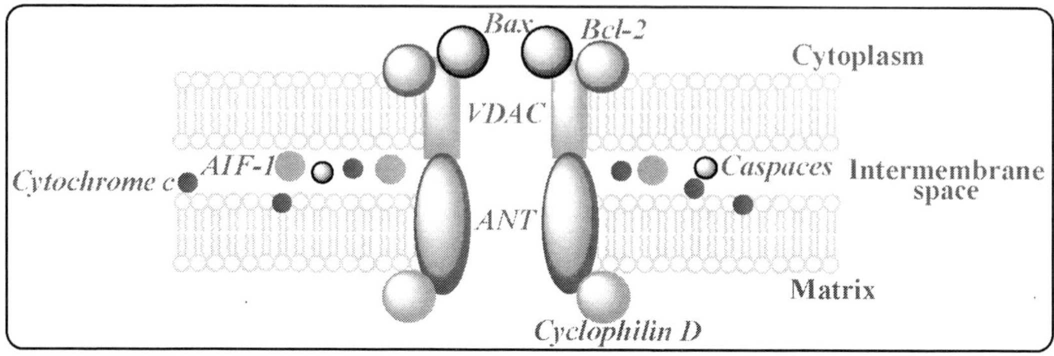


Figure 37. The mitochondrial permeability transition pore communicating the mitochondrial matrix and cytosol through a junction formed between the inner and the outer mitochondrial membranes. It is formed mainly of the inner membrane adenine nucleotide translocase associating cyclophilin, and, the outer membrane voltage-dependant anion channel (VDAC) associating either Bcl-2 or Bax. Its formation gives an escape for inter-membranous and inner membrane proteins (caspases, cytochrome c and AIF-1 = apoptosis inducing factor-1).

*Worth Noting: Natural measures against a destructive role for mitochondria*

- *Uncoupling proteins:* They dissipate the proton gradient and prevent free radical generation. The expression of natural uncoupling proteins declines with aging.
- *Autophagy by Lysosomes:* Efficiency of autophagy of the damaged mitochondria by lysosome declines with age, resulting in more mitochondria producing higher levels of superoxide and lower energy.
- *Antioxidants*: The oxidative stress-inducible mitochondrial Mn-superoxide dismutase; and catalase in heart mitochondria decline with age.
- *Cardiolipin*: Cardiolipin sequesters cytochrome c in the inner mitochondrial membrane. Both decline significantly with aging.

## Drug-induced Mitochondrial Toxicity

Drug-induced mitochondrial toxicities induce symptoms resembling those of the Oxphos diseases; myopathies (skeletal and cardiac), neuropathies (peripheral and central), pancreatitis, lactic acidosis, hepatic steatosis, bone marrow suppression and lipodystrophy. These drugs include; the aforementioned antibiotics, the nucleoside reverse transcriptase antiretroviral inhibitors, analgesics (e.g., acetaminophen and nimesulide), antipsychotics (e.g., barbiturates, amitriptyline, chlorpromazine, valporic acid, tolcapone and lithium), the lipid-lowering Statins HMG-CoA reductase inhibitors, alcohol and other socially additive drugs. They act through; a direct enzyme inhibition (metabolic enzyme, the mtDNA polymerase $\gamma$ and transcription), inducing mtDNA mutations and depletion, lower energy production, inhibition of $CoQ_{10}$ and cholesterol synthesis (as the case for Statins) and inducing mitochondrial permeability and oxidative stress (consumption of antioxidants and increasing free radical generation) that culminate into cell death by apoptosis. This necessitates mitochondrial toxicity testing for new drugs to be approved for human use. Mostly these tests depend on measuring the mitochondrial enzyme activity in a skeletal muscle biopsy.

## Review Questions

*A. Write short note on:*

1. Basic features of the mitochondrial DNA.
2. Human metaphase chromosome.
3. DNA fingerprints.
4. B-form of DNA.
5. The role to topoisomerases in DNA replication.
6. DNA supercoiling.
7. Origin of replication complex.
8. Inhibitors of DNA synthesis.
9. Functions of the telomere.
10. Fates of point mutation.
11. The positive and the negative regulators of the cell cycle.
12. The p53.
13. The direct reversal for repair of DNA mutations.
14. Mutations repaired by the base excision repair.
15. Inhibitors of the cell cycle.
16. Structural chromosomal abnormalities.
17. Numerical chromosomal abnormalities.
18. Basic technique of traditional cytogenetics.
19. Role of mitochondria in cell death.
20. The transcription-coupled DNA repair.

*B. Explain the following:*

1. Coordination of DNA replication and the cell cycle.
2. Suppressor mutations.
3. Telomere synthesis.
4. Mutagen/carcinogen/teratogen.
5. Effect of DNA mutation.
6. Role of the NAD-dependent poly(ADP-ribose) polymerase.
7. The transcription-coupled nucleotide excision repair.
8. Diseases related to nucleotide excision repair.
9. The Meselson-Stahl experiment for semi-conservative nature of DNA replication.
10. The cell cycle check-points.
11. Action of the different inhibitors of the cell cycle.
12. The telomere role in aging.
13. The mitochondrial role in aging.
14. Bases of the mitochondrial toxicity tests.
15. Mitochondrial DNA related diseases affect active tissues in a maternally inherited manner.

16. Nucleosome is a dynamic unit of the chromatin.
17. Stained chromosome banding.
18. Hetero- vs. euchromatin.
19. Mitochondrial DNA is prone to mutation relative to the nuclear DNA.
20. OXPHOS diseases are manifested by .........

C. *True and False Questions:*

1. A mutagen in most of cases a carcinogen.
2. A teratogen is a mutagen or a non-mutagen chemical that causes embryonic developmental abnormalities.
3. Non-sense mutation generates longer protein, whereas, read-through mutation generates shorter protein.
4. Frame shift mutation is milder than a silent mutation.

D. *Supply the Missing Information Questions:*

1. Chromatin is formed of:
   a. Double stranded chromosomal ----------.
   b. ------------ basic proteins.
   c. Non-histone structural and regulatory -----------------.
2. ------------.
3. Types of chromatin include:
   a. The transcriptionally-inactive constitutive -----------------------.
   b. The transcriptionally-active --------------------.
   c. The transitional ---------------- heterochromatin.
   4. Types of DNA sequences:
   a. The unique non-repetitive --------------------- sequences represent 1.1% of the human genome
   b. The gene-related regulatory and ------------------------ sequences represent 42% of the human genome.
   c. ----------------------------- sequences constitute more than 50% of the human genome.
5. Minisatellites are intergenic repetitive sequences that have two types; ------------ --------------------------------- and --------------------------.
6. Minisatellites short tandem repeats segregate independently and are heritable in a Mendelian fashion, therefore, can be used for forensic individual-specific -------- ----------------------------.
7. Mitochondria are acquired into the daughter cells mainly from --------------------- ---.
8. The mitochondrial DNA encodes for a few of the mitochondrial -----------------.
9. The mitochondrial DNA replication occurs semi-autonomous during ------------- --- of cell cycle.
10. The mitochondrial protein synthesis resembles the prokaryotic system in its sensitivity to inhibition by ----------------.
11. Mutations in mitochondrial genes and related diseases have --------------------- of non-Mendelian inheritance.
12. Analysis of mitochondrial DNA is used for forensic determination of ------------- ----- relatives.

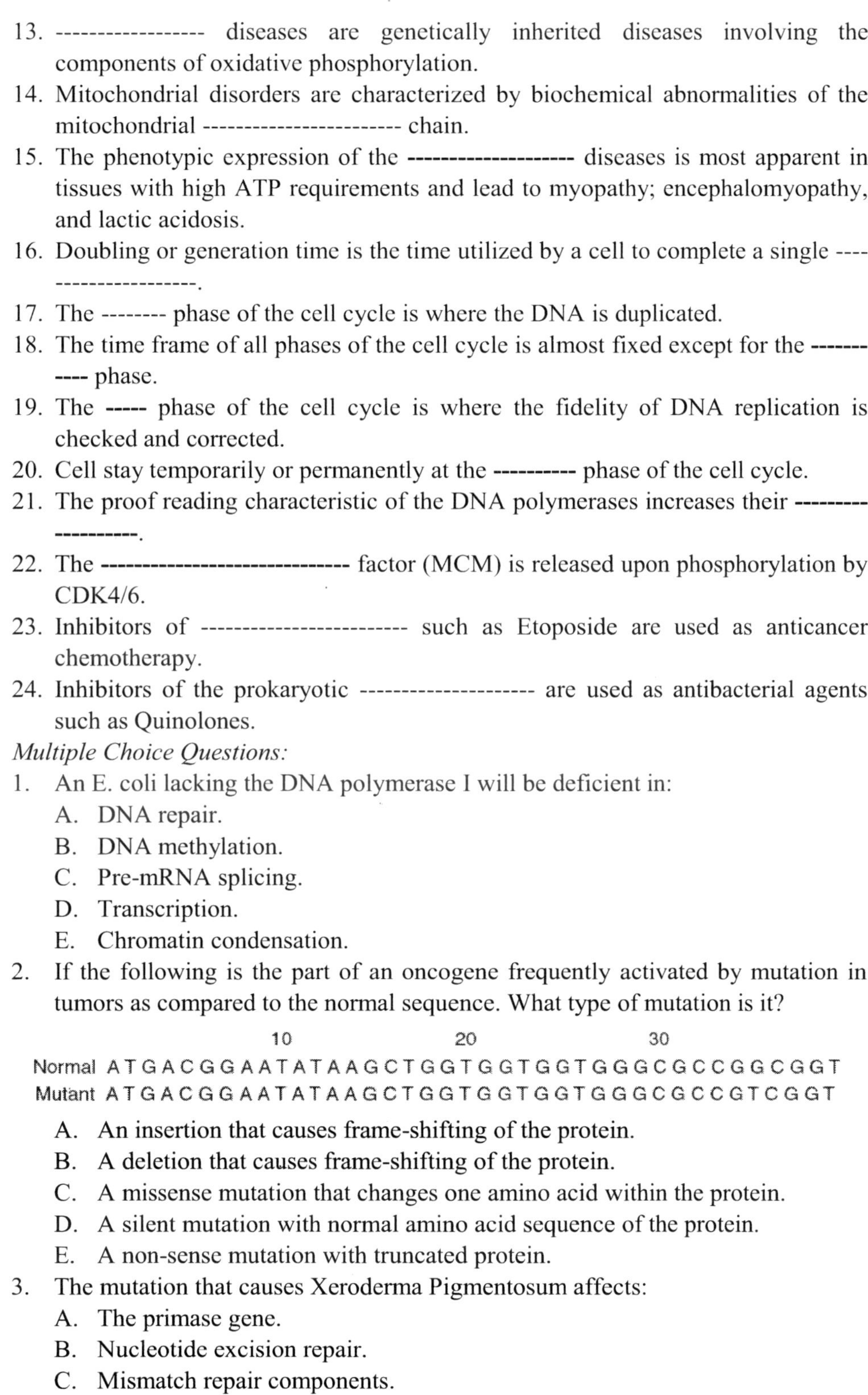

13. ----------------- diseases are genetically inherited diseases involving the components of oxidative phosphorylation.
14. Mitochondrial disorders are characterized by biochemical abnormalities of the mitochondrial ----------------------- chain.
15. The phenotypic expression of the ------------------- diseases is most apparent in tissues with high ATP requirements and lead to myopathy; encephalomyopathy, and lactic acidosis.
16. Doubling or generation time is the time utilized by a cell to complete a single --------------------.
17. The -------- phase of the cell cycle is where the DNA is duplicated.
18. The time frame of all phases of the cell cycle is almost fixed except for the ---------- phase.
19. The ----- phase of the cell cycle is where the fidelity of DNA replication is checked and corrected.
20. Cell stay temporarily or permanently at the ---------- phase of the cell cycle.
21. The proof reading characteristic of the DNA polymerases increases their -------------------.
22. The ---------------------------- factor (MCM) is released upon phosphorylation by CDK4/6.
23. Inhibitors of ----------------------- such as Etoposide are used as anticancer chemotherapy.
24. Inhibitors of the prokaryotic -------------------- are used as antibacterial agents such as Quinolones.

E. *Multiple Choice Questions:*

1. An E. coli lacking the DNA polymerase I will be deficient in:
   A. DNA repair.
   B. DNA methylation.
   C. Pre-mRNA splicing.
   D. Transcription.
   E. Chromatin condensation.
2. If the following is the part of an oncogene frequently activated by mutation in tumors as compared to the normal sequence. What type of mutation is it?

```
                 10                  20                  30
Normal ATGACGGAATATAAGCTGGTGGTGGTGGGCGCCGGCGGT
Mutant ATGACGGAATATAAGCTGGTGGTGGTGGGCGCCGTCGGT
```

   A. An insertion that causes frame-shifting of the protein.
   B. A deletion that causes frame-shifting of the protein.
   C. A missense mutation that changes one amino acid within the protein.
   D. A silent mutation with normal amino acid sequence of the protein.
   E. A non-sense mutation with truncated protein.
3. The mutation that causes Xeroderma Pigmentosum affects:
   A. The primase gene.
   B. Nucleotide excision repair.
   C. Mismatch repair components.
   D. Synthesis DNA across the damaged region.
   E. Proofreading capacity.

4. All of the following statements are correct about cyclin B, EXCEPT:
   A. It is synthesized anew in every cell cycle.
   B. It activates cdk2.
   C. It is present in the cell during G2 phase.
   D. It is degraded via ubiquitin pathway.
   E. It is required for the cell to exit from the mitotic phase.
5. A DNA polymerase contains a lysine residue that is important for its ability to bind DNA. A mutation converted lysine into the following pairs of amino acids. Which pair of them would be the most and the least harmful to the ability to bind DNA?

| | Most | Least |
|---|---|---|
| A. | Valine | Aspartate. |
| B. | Valine | Arginine. |
| C. | Arginine | Glycine. |
| D. | Glutamate | Valine. |
| E. | Glutamate | Arginine. |

6. The specialized structures located at the ends of the eukaryotic chromosomes are called:
   A. Terminators.
   B. Telomeres.
   C. Terminal un-translated sequence.
   D. Translation stop signal.
   E. TATA box.
7. For DNA polymerase I, Which of following serves as a substrate for DNA synthesis?
   A. A single-stranded circular DNA.
   B. A single-stranded circular DNA that is base-paired to a small linear DNA fragment with a free 3'-OH.
   C. A single-stranded circular DNA that is base-paired to a small linear DNA fragment with a free 3'-phosphate.
   D. A double-stranded circular DNA.
   E. A blunt-ended double-stranded linear DNA with free 3'-OH on both ends.
8. After completion of S-phase of a mammalian cell cycle all are correct, EXCEPT:
   A. Histone content is double its amount at G1.
   B. Bases of the new strand are paired to those of the parent strand.
   C. Each chromosome has four telomeres.
   D. The nucleus contains tetraploid amount of DNA.
   E. Sister chromatids are disjoined.
9. An accurate DNA replication requires all of the following EXCEPT:
   A. An RNA primer.
   B. DNA ligase.
   C. A free 5'-OH.
   D. The 3'⇒5' exonuclease activity.
   E. Topoisomerase activity.
10. Which of the following is true about DNA gyrase?
   A. It unwinds DNA.

B. It is found in both prokaryotes and eukaryotes.
C. It is similar to type I topoisomerase.
D. It produces positive supercoils.
E. It is energy independent.

11. A common lesion found in DNA after exposure to UV light:
    A. Pyrimidine dimers.
    B. Single strand breaks.
    C. Base deletion.
    D. Purine dimers.
    E. Frame-shift mutation.
12. One of the following is most probably to cause loss of gene function:
    A. A missense mutation in the coding sequence.
    B. A change from TAA into TAG in the coding sequence.
    C. A frame-shift mutation at the end of the coding sequence.
    D. A mutation in the 3'-untranslated sequence.
    E. A T to C mutation in the promoter conserved box sequence.
13. A mammalian cell with a doubling time of 24 hrs was synchronized at G1 and labeled by bromodeoxiuridine that increases the density of the synthesized DNA for two days. The isolated chromosomal DNA was analyzed by density-gradient sedimentation. Which of the following patterns of DNA would be expected?
    A. 100% heavy DNA.
    B. 100% light/heavy DNA.
    C. 50% heavy/heavy and 50% light/heavy DNA.
    D. 50% heavy/heavy and 50% light/light DNA.
    E. 25% heavy/heavy; 50% light/heavy; 25 light/light DNA.
14. Separation of DNA strands during replication requires:
    A. Helicase.
    B. 3'-5'-exonuclease.
    C. DNA ligase.
    D. Primase.
    E. RNaseH.
15. Events happening in the G1 phase of the cell cycle include ALL except:
    A. A reduction in the size of the cell.
    B. Cell commitment to cell division.
    C. Replication of the cytoplasmic organelles particularly mitochondria.
    D. Expression and storage of materials required for DNA replication.
16. Topoisomerases are required for:
    A. Separation of the DNA double strands.
    B. Prevention of the reassociation of the two single stranded DNA.
    C. DNA recombination.
    D. Removal of the DNA supercoils.
17. Considering the telomere - all are CORRECT except:
    A. Is a double stranded non-coding gene-less sequence.
    B. Locates to the centromeric area of the chromosome.
    C. Is maintained by telomeric repeat-binding factors.
    D. Is synthesized by telomerase.

18. Considering the telomere - all are CORRECT except:
    A. Is the complementary sequence to the telomeric RNA.
    B. Is synthesized by a reverse transcriptase.
    C. Activated joining among chromosomes.
    D. Its length is longest in the zygote, germ, young and cancer cells.
19. All of the following enzymes have proofreading characteristic EXCEPT:
    A. Ribosome.
    B. DNA polymerases.
    C. Amino-acyl-tRNA synthetases.
    D. Reverse transcriptase.
20. The silent or synonymous mutation:
    A. Reflects the degeneracy of the genetic code.
    B. Has a change at the $1^{st}$ base of the codon.
    C. Generates a new amino acid.
    D. Locates to euchromatin.
21. The following apply to missense or non-synonymous mutation EXCEPT:
    A. Occurs either in $1^{st}$ or $2^{nd}$ base of the codon.
    B. Generates a different amino acid.
    C. Is acceptable at varying degrees.
    D. Could result in a stop codon and a truncated protein.
22. Considering missense mutation:
    A. Several types of hemoglobin are normal due to an acceptable missense mutation.
    B. Sickle cell anemia or hemoglobin S is an example of partially acceptable missense mutation.
    C. Hydrophilic amino acid replacement for hydrophobic one is more damaging.
    D. Unacceptable missense is exemplified by methemoglobinemia (HbM or Hb Boston).
23. Considering palindromes or palindromic nucleic acid sequences - all are CORRECT except:
    A. They are complementary sequences readable the same in opposite directions.
    B. Targets for most of the restriction endonucleases.
    C. Important determinant of the 3D structure of the DNA or RNA molecules.
    D. Are nuclear scaffold binding sites.
24. Considering palindromes or palindromic nucleic acid sequences - all are CORRECT except:
    A. Are targets for binding of regulatory proteins that affect rate of gene expression.
    B. Are exon-intron junction sequences.
    C. Strongly influence the 3D structure of the nucleic acid molecules.
    D. Those of RNA bind complementarily specific gene sequences to inhibit gene expression.
25. Nucleosome as the packaging unit of the chromatin is formed of ALL except:
    A. Core histone (two each of H2A, H2B, H3 and H4).
    B. 1.75 superhelical turns (146 nucleotides length) of DNA wrapped around the core histone.

C. A linker stretch of DNA (30 nucleotide in length) connecting two nucleosomes.
D. A ribosome.

26. Nucleosome structure has several dynamic regulatory roles include ALL except:
    A. Control of rate of gene expression (induction/repression and silencing).
    B. Control of chromosomal condensation (hetero- vs. euchromatin) and assembly.
    C. DNA repair.
    D. Elongation of the telomere.
27. Considering mitochondrial DNA:
    A. Is formed of a few molecules of doubled stranded circular DNA.
    B. Is maternally inherited.
    C. is more vulnerable to mutations than nuclear DNA.
    D. Resembles the prokaryotic system.
    E. All of the above.
    F. None of the above.
28. In a double stranded DNA if the cytosine content is 20% of the total bases, the adenosine content will be:
    A. 10%.
    B. 20 %.
    C. 30 %.
    D. 40 %.
    E. 80%.
29. The DNA strand is composed of:
    A. Sugars and bases.
    B. Phosphates and deoxyribonucleosides.
    C. Phosphates and sugars.
    D. Nucleotides and sugars.
    E. Phosphates and ribonucleosides.
30. In a circular double stranded DNA molecule with 21% adenosine, the guanosine % is:
    A. 10.5.
    B. 21.
    C. 29.
    D. 58.
    E. No enough information for calculation.
31. All of the following are correct for histone proteins except:
    A. Present in prokaryotes and eukaryotes.
    B. Basic positively charged proteins.
    C. Are not susceptible to covalent modifications.
    D. Carry an inert packaging role for DNA.
    E. Bind to mitochondrial DNA.
32. Similarity between prokaryotic system and mitochondria include all, EXCEPT:
    A. Inhibitors of transcription.
    B. Having double stranded circular DNA.
    C. Inhibitors of protein synthesis.

D. Ability of both to synthesis all protein they need.
E. Absence of repetitive and intron sequences.

33. For the 5'-ATCGATCGATCGATCG-3' DNA sequence, which of the following is the complementary strand in what direction?
    A. 5'-ATCTATCGATCGATCG-3'.
    B. 3'-ATCGATCGATCGATCG-5'.
    C. 5'-CGAUCGAUCAUCGAU-3'.
    D. 5'-CGATCGATCGATCGAT-3'.
    E. 3'-CGATCGATCGATCGAT-5'.
34. The backbone of a DNA strand is composed of which of the following?
    A. Sugars and bases
    B. Phosphates and sugars
    C. Bases and phosphates
    D. Nucleotides and sugars
    E. Phosphates and nucleosides
35. With a polymerization rate of 1 kb per second, a 1,000-kb double stranded DNA with 10 evenly spaced origins of replication would be replicated in;
    A. 20 seconds.
    B. 30 seconds.
    C. 40 seconds.
    D. 50 seconds.
    E. 100 seconds.
36. The primase is not required during DNA repair processes because of which of the following?
    A. The primase only associates the replication complex.
    B. The RNA primer would increase the natural mutation rate.
    C. The primer is not required for the DNA polymerase I.
    D. The primer is not required for the DNA polymerase III.
    E. The repair enzymes use the free DNA 3-OH for replacing the defective sequence.
37. The separation of DNA strands during replication is activated by;
    A. Helicase.
    B. An exonuclease.
    C. DNA ligase.
    D. Primase.
    E. An endonuclease.
38. Xeroderma pigmentosum involves defective;
    A. Primase gene.
    B. UV-damaged DNA repair.
    C. Mismatch repair components.
    D. DNA polymerase I.
    E. Loss of the DNA proofreading ability.
39. Inactive genes are associated with;
    A. Hyperacetylated opened chromatin.
    B. Deacetylated condensed chromatin.
    C. Highly phosphorylated chromatin.

D. Deacetylated condensed chromatin.
E. Closed hyperacetylated chromatin.

40. DNA content of normal mammalian cell following DNA replication and prior cell division is;
    A. Sub-diploid.
    B. Polyploid.
    C. Tetraploid.
    D. Mixoploid.
    E. Haploid.
41. Which cyclin regulates CDK4/6 catalytic activity?
    A. Cyclin A.
    B. Cyclin B.
    C. Cyclin C.
    D. Cyclin D.
    E. Cyclin E.

*Answer key for the True and False Questions:* 1, T; 2, T; 3, F; 4, F.

*Answer key for the Answer key for the Supply the Missing Information Questions:* 1, DNA, Histone, proteins, RNA; 2, heterochromatin, euchromatin, facultative; 3, single-copy gene exon, intervening intron, intergenic repetitive; 4, variable number of tandem repeats, short tandem repeat; 5, DNA typing or fingerprinting; 6, the mother ovum; 7, proteins; 8, G1 phase; 9, antibiotics; 10, a maternal pattern; 11, motherhood; 12, Oxphos; 13, Respiratory; 14, Oxphos; 15, cell cycle; 16, S; 17, G1; 18, G2; 19, Go; 20, fidelity; 21, replication licensing; 22, topoisomerases; 23, DNA gyrase.

*Answer key for the MCQs*: 1, A; 2, C; 3, B; 4, E; 5, E; 6, B; 7, B; 8, E; 9, C; 10, A; 11, A; 12, E; 13, C; 14, A; 15, A; 16, D; 17, B; 18, C; 19, D; 20, A; 21, D; 22, C; 23, D; 24, B; 25, D; 26, D; 27, E; 28, D; 29, B; 30, C; 31, B; 32, D; 33, D; 34, B; 35, D; 36, E; 37, A; 38, B; 39, A; 40, C; 41, D.

*Chapter III*

# Ribonucleic Acids (RNAs) and Protein Synthesis: The Complete Story of the Central Dogma

## Topics Discussed

- Structure and types of RNA
- Flow of genetic information and central dogma
- Gene concept
- RNA synthesis (Transcription) and post-transcriptional processing of RNA and related diseases
- Regulation of prokaryotic and eukaryotic gene expression
- Genetic code and protein synthesis (Translation)
- Post-translational modifications and protein targeting and related diseases.

## Learning Objectives

After understanding this part the student should be able to:

- Identify the major types of cellular RNA and the function of each
- Describe the major steps in transcription of an RNA molecule
- Explain the function of the different RNA polymerase enzymes
- Appreciate the complexicity in controlling rate of transcription and the many interacting proteins required
- Describe the different processing and splicing events that occur during synthesis of eukaryotic mRNAs and bases of related diseases
- Describe the general mechanisms of regulation of gene expression, including the role of cis-acting promoters and enhancers and trans-acting transcription factors

- Describe the many levels at which gene expression may be controlled, using phosphoenol pyruvate carboxykinase gene expression as a model
- Explain how alternate splicing, alternate promoters and post-transcriptional editing can modulate expression of a gene
- Explain how the structure and packaging of chromatin can affect gene expression
- Explain how genomic imprinting can result in changes in gene expression, depending on whether alleles are maternally or paternally inherited
- Outline the structure and redundancy of the genetic code
- Describe the major steps in translation of an mRNA into polypeptide through interaction of different RNAs
- Describe the different posttranslational processing, targeting, and degradation of proteins and bases of related diseases.

# Structure and Types of RNA

RNA contains ribose instead of deoxyribose as the sugar moiety. RNA is very unstable compared to DNA and is a single stranded polymer of ribonucleotides of adenine, guanine, cytosine and uracil. It is found in nucleus, nucleolus, cytoplasm, mitochondria and the genome of RNA viruses (retroviruses).

A single mammalian cell contains a total of approximately $1x10^{-5}$ µg of RNA; 80 - 85% is ribosomal RNA (rRNA); 15 - 20% is composed of a variety of low molecular weight species, i.e., transfer RNA (tRNA) and small nuclear RNA (SnRNAs); and 1 - 5% messenger RNA (mRNA). The rRNAs and tRNAs are of defined size and sequence and therefore relatively homogenous. Oppositely, mRNA is highly heterogeneous with respect to; size (~100 - 14,000 nucleotides), half-life and sequence, and hence, is called heteronuclear RNA (hnRNA). Other than these main types of RNA, there are other minor but essential types including the catalytic ribozymes; see below for details.

*The different forms of RNA differ in* gene of origin, function and location, size, posttranscriptional structural modification and half-life. Although RNAs are single stranded, intrachain base complementation acquired RNAs secondary and tertiary structures, e.g., hairpins, stem-loops and pseudoknots that may specify structural domains.

## 1. Messenger RNA (mRNA)

Messenger RNA is the single stranded coding nucleic acid that transfers a copy of the genetic code from the nucleus to the cytosol for proteins synthesis. Actually they carry the transcript of the nucleotide sequences corresponding to the sequence of amino acids in the protein they code for. Therefore, there are thousands of types of specific mRNAs at least one for every polypeptide-coding gene (some genes give several mRNAs).

The mRNA is highly heterogeneous with respect to size (~100 – 14,000 nucleotides); half-life/stability and sequence except for the 3'-terminus, which in eukaryotic cells mostly carries a poly-adenylic acid (polyA) sequence. This is why the initial single-stranded blue scripts and intermediate processing products of mRNAs are called *heterogeneous nuclear*

*RNAs (hnRNAs or pre-mRNAs)*. These thousands of types of intermediary mRNA forms constitutes less than 0.1% of the total cellular RNA and ranges from 0.5 – 500 kb in size. Thus, these intermediates may be 5-10 times larger than their mature mRNA forms.

The mRNA is the copy (complementary image) of the base sequence of the antisense-template strand of the gene proper in the nucleus. Therefore, mRNA carries the genetic information (code, single word is a codon) for protein synthesis. Each amino acid is represented by three consecutive nucleotides sequence, i.e., the codon that is recognized by the anticodon of the tRNA.

The mRNA is characterized in eukaryotes by having extensive post-transcriptional modifications viz. (see Figure 38):

- The translation-non-functional intervening sequences, i.e., the *introns* are removed.
- *A polyadenylate tail* (up to 250 bases) is added at the 3'-end. The polyA tail marks the mRNA for transportation to the cytoplasm, control rate of its translation and protects it from degradation by 3'-exonuclease. The polyA tail is used to purify mRNA from other RNAs by attachment to a stationary phase oligo-deoxythymidine polymers.
- *7-methyl-guanosine triphosphate-5'-5'- cap* is added at the 5'-end of mRNA. This cap directs the initiation of translation and protects the mRNA from degradation by 5'-exonuclease.

  After all of this processing, mature mRNA still has 5'- and 3'-terminal un-translated regions (UTR). These regions contain sequences that control rate of mRNA translation and mRNA half-life.

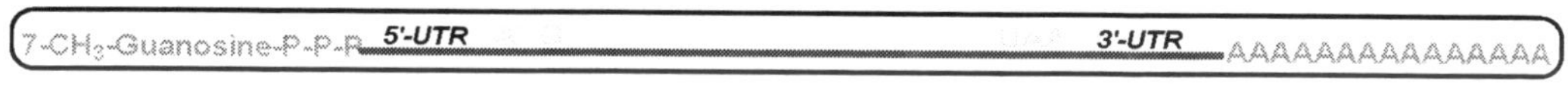

Figure 38. Diagrammatic structure illustrating the main features of the mature mRNA. UTR = un-translated region.

## 2. Transfer RNA (tRNA)

The tRNA is the simplest and smallest non-coding RNA, having only 74 - 95 nucleotides length and constituting ~5% of the total cellular RNA. Like mRNA, tRNA is transcribed from specific genes in the nucleus – there are >40 genes dispersed through the genome with several copies of each type of tRNA gene as well as tRNA pseudogenes. The post-transcriptional modifications of tRNA include; unit cutting from a large transcript, removal of introns, extensive modifications of its bases, and addition of the amino acid acceptor nucleotide sequence 3'-ACC-5' at the 3'-end of the acceptor arm (see Figure 39)

The structure of tRNA is characterized by the presence of several intra-chain hydrogen bonds that give it a clover-leaf like shape. The molecule is folded to form four loops and the 3'-ACC sequence viz.:

a) The D loop and arm: The loop is named so because it contains an unusual nitrogen base, i.e., *dihydrouridine* (DHU). The loop is important for proper recognition of a given tRNA by its *amino-acyl-tRNA synthetase*.

b) The TΨC loop and arm: The loop contains an unusual uracil nucleotide known as *pseudouridine* (Ψ) with a carbon-carbon bond between C1 of the sugar and C5 of the base due to rearrangement of a preformed UMP. The 'Thymine-Pseudouridine-Cytosine' sequence is characteristics to the loop. The arm is involved in *binding of amino-acyl-tRNA to ribosome*.
c) Anticodon arm: The loop carries the 'anticodon' triplet nucleotide sequence that recognizes as specific codon among the '*Code' on the mRNA molecule*.
d) The variable loop: This is an extra loop hanging between TΨC and the anticodon arms and varies in size from one tRNA to the other.
e) The 3'-ACC-amino acid acceptor end: It carries a specific amino acid to be transferred into the ribosomes for protein synthesis. Therefore, each amino acid has at least one tRNA. There are 33 types of tRNA in cytoplasm and only 22 types in mitochondria.

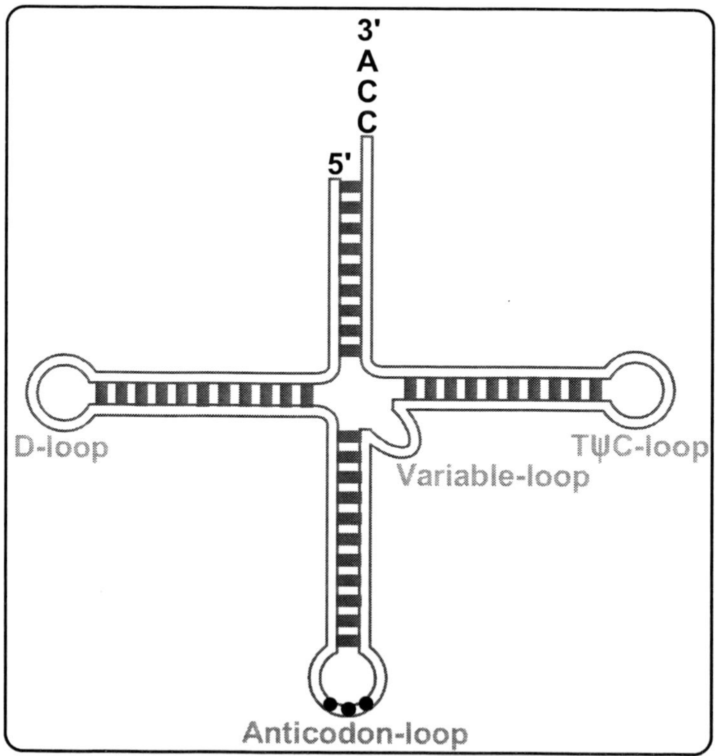


Figure 39. Diagrammatic structure illustrating the main features of the mature tRNA.

## 3. Ribosomal RNA (rRNA)

The rRNAs are the types of non-coding RNA found in the ribosomes, in association with the proteins. The nucleoprotein assembly acts as the proteins synthetic machinery in the cytosol. Polysomes are assemblies of several ribosomes connected to one mRNA ~100 nucleotide apart. The ribosomal mass constitutes ~40% of the total cell mass.

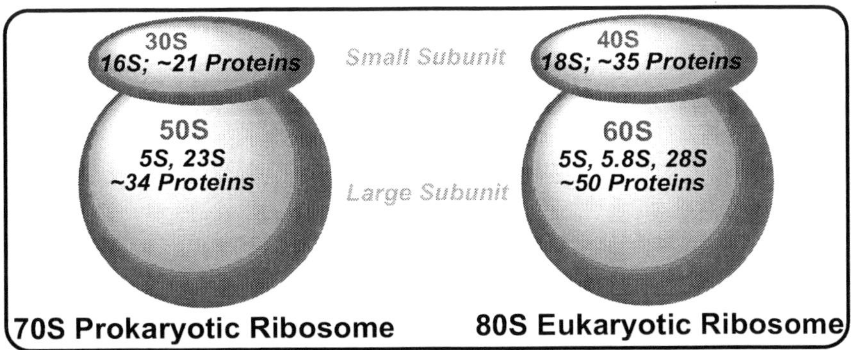


Figure 40 Diagrammatic structure illustrating the main features of the prokaryotic vs. the eukaryotic ribosome and the size and content of each of their subunits.

These rRNA are synthesized from a large number (repeated 50X in tandem) of rRNA genes on the antennae (the very short p arm) of 5 chromosomes number; 13, 14, 15, 21 and 22, ramified in the nucleolus. The 5S rRNA has multiple repeats in 3 clusters on chromosome 1. The initial large transcript is cut into the smaller molecules that are subsequently subject to extensive chemical modifications onto their nucleotides. The highly methylated rRNA molecules are assembled in the nucleolus with specific ribosomal proteins into ribosomes.

The rRNAs constitutes ~80% of the total RNA content. They are classified according to their sedimentation coefficient (depended on density and shape of molecules) during ultracentrifugation on NaCl solution in Svedberg units (S) into; 5S, 16S and 23S in prokaryotes and mitochondria (that are; 120, 1700, and 3700 nucleotides, respectively). The 5S, 5.8S, 18S and 28S rRNAs exist in eukaryotes (that are; 120, 160, 1900 and 4800 nucleotides in length, respectively). Some of these rRNAs have enzyme activity, i.e., are ribozymes.

Eukaryotic ribosome is 80S in size and is formed of two subunits; 60S and 40S. 60S subunit is composed of ~50 protein molecules and 5S, 5.8S and 28S rRNAs. The 40S subunit is composed of ~35 protein molecules and 18S rRNA. Prokaryotic ribosome is 70S in size and formed of 50S and 30S subunits. The 50S is formed of ~34 protein molecules and 23S and 5S rRNA. The 30S subunit is formed of ~21 proteins and 16S rRNA (see Figure 40). It is similar to the mitochondrial ribosome that is 55S.

## 4. Small/Stable Nuclear RNA (snRNA)

They are large number (>200 type) of non-coding small discrete highly conserved and stable RNA species, found in nucleus and/or cytoplasm complexed in ribonucleoprotein complexes. They range from 90 – 300 nucleotides and are present at 100,000 – 1,000,000 copies per cell constituting less than 1% of the total cellular RNA content. SnRNAs are a subset that is involved in mRNA processing and regulation of gene expression with about 30 types. Examples are the U1, 2, 4, 5, and 6 snRNAs involved in introns removal in spliceosome. Some of these snRNAs are ribozymes. U4 and 6 may also be required for polyA tail processing. U7 may be involved in production of the correct 3'-end of histone mRNA that lacks polyA tail.

## 5. Small Nucleolar RNAs (snoRNAs)

They are similar to the snRNAs but function mainly in the nucleolar organizer area for rRNA posttranscriptional modifications including methylation and conversion of uridine into pseudouridine. They may have more than 200 types and their size range is 70-100 nucleotides constituting less than 1% of the total cellular RNA content. They are transcribed from specific genes and from introns of other genes.

## 6. Small Cytoplasmic RNAs (scRNAs)

They are tens of types of non-coding small RNAs (90-330 nucleotides in size) constituting less than 1% of the total cellular RNA content that are located in the cytoplasm. They function in protein synthesis and transport mainly as a major part of the ***signal recognition particle*** that temporarily interrupts protein synthesis to direct the nescient polypeptide sequence into endoplasmic reticulum after which the synthesis resumes.

## 7. Telomeric RNA (telRNA)

It is one type of non-coding RNA component in the ribonucleo-protein reverse transcriptase (RNA-dependent DNA polymerase) complex responsible for synthesis and maintenance of the chromosome ends, i.e., the telomere. Telomeres stabilize the ends of chromosomes and determine the replicative ability of cells. The telRNA molecule functions as a template for its own enzyme to synthesize species-specific telomere. Human telRNA is 450 nucleotides in size out of which 5′-CCCUAA-3′, is reverse transcribed into the 5′-TTAGGG-3′ telomeric DNA repeat. It constitutes very small % of the total cellular RNA content.

*Worth Noting: Other very important types of non-coding RNAs*

They include; the 25 kb non-coding *Xist RNA* that is transcribed from Xist gene at the X chromosome inactivation center. It forms a ribonucleo-protein complex that inactivates one of the two X chromosomes into inactive heterochromatin conformation. *RNases* acting on processing of pre-tRNAs and some pre-rRNAs are ribonucleoprotein complexes. Example is *RNase P* (a holoenzyme formed of one RNA molecule and at least one protein molecule) catalyzes specific phosphodiester bond hydrolysis in pre-tRNAs to produce mature tRNA 5'-ends. *RNA primers* synthesized by DNA-dependent RNA polymerase - the primase - are essential for initiation of DNA replication, where, DNA polymerases cannot start a *de novo* polymerization reaction. One type of the natural post-transcriptional gene silencing is executed by *small-* and *micro-interfering RNAs* (siRNAs and miRNAs) produced endogenously in human and other organisms, e.g., Neurospora crassa. Establishment of double-stranded RNA (dsRNA) post-transcriptionally between a specific mRNA and siRNA/miRNA suppresses the mRNA expression, a process known as RNA interference (RNAi). They activate hydrolysis of the mRNA by RNaseH, its decapping and sterically hinder its translation. These siRNAs have several functions such as differential tissue-specific gene expression, antiviral defense, chromatin remodeling and stabilizing the genome. Other types of RNAs function in control of rate of transcription, editing, translation and stability and degradation of RNA. The group of viruses called retroviruses contains RNA as the genomic nucleic acid.

# The Flow of Genetic Information

The genome expression from DNA to RNA to protein and function necessitates several regulated steps that constitute the major components of the central dogma to enable cells to express a specific subset of genes at specific tissue, time and conditions and could be summarized in the following:

- *Gene accessibility* necessitates that the active gene(s) in a cell being accessible by the transcription initiation complex and not to be a heterochromatin component.
- *Transcription of the gene into RNA* and *posttranscriptional processing of RNA* into the mature functional form
- *Control of the RNA half-life through storage and/or degradation.*
- *RNA translation into a polypeptide*
- *The posttranslational processing* of the polypeptide/protein into final functional three-dimensional configuration with proper proteolysis, covalent modifications, folding and aggregation into protein
- *Control of the protein half-life through degradation.*

Such flow of genetic information in one direction from the stored gene code in DNA by transcription into a message copy as mRNA that is then translated into a protein that executes the gene function occurs in almost all organisms and was coined the name of '*Central dogma*' by *Francis Crick* 1958; see Figure 41 for a summary of the major landmarks of central dogma.

Nowadays, central dogma has exceptions because in a few cases the information may flow from RNA to DNA then to RNA then to proteins. The example is retroviruses with single stranded RNA (ssRNA) as the genomic molecule that necessitates reversal of the flow of the genetic information from RNA into DNA then into mRNA then protein in the life cycle of these viruses.

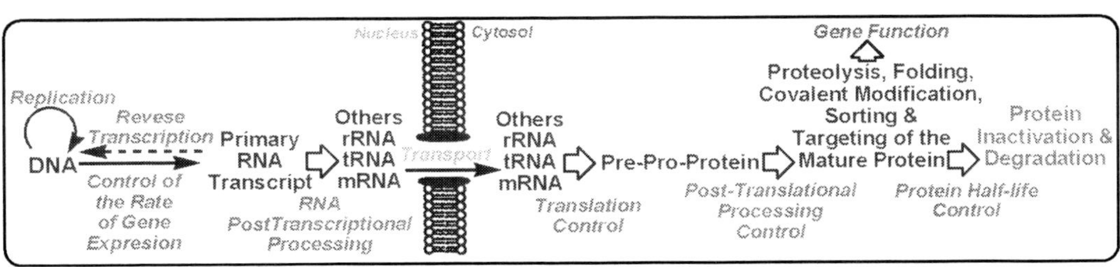


Figure 41. The major control stations composing the central dogma.

These viruses contain an enzyme named *Reverse transcriptase (RT)* that is an RNA-dependent DNA polymerase. Although retroviral reverse transcriptases are structurally diverse, they all possess three functional activities namely; RNA-directed DNA polymerase activity to create a single stranded complementary DNA (sscDNA) from the viral RNA; RNaseH activity to digest the viral genomic RNA strand; and DNA-directed DNA polymerase activity that synthesizes complementary DNA strand to the sscDNA to make the proviral dscDNA - using the RT-created extranucleotide DNA overhung as a primer. Note that cDNA is always unnatural, i.e., made and there is no organism with a natural cDNA as a genome, whereas, genomic DNA means the naturally occurring DNA.

The reverse transcriptase is targeted for inhibition by the nucleotide reverse transcriptase inhibitors for treatment of HIV-AIDS and different types of retroviral influenzae (avian and swine). These reverse transcriptase inhibitors are dideoxynucleosides/nucleotides that are incorporated by RT into the viral DNA (after intracellularly activation into their triphosphorylated nucleotide form) leading to DNA chain termination. Examples of these modified nucleoside/nucleotides RT inhibitors include; Zidovudine (AZT or AZD), Abacavir (ABC), Didanosine (ddI), Emtricitabine (FTC), Lamivudine (3TC), Stavudine (d4T), Tenofovir (TDF), Zalcitabine (ddC), and Emtricitabine (Emtriva). The regular DNA polymerase has a lower affinity to the dideoxynucleotides than the RT. However, they are also incorporated into the human mitochondrial DNA that gives rise to some toxic side-effects, e.g., hepatotoxicity. Zidovudine - azidothymidine - was the first drug approved for HIV-AIDS treatment and is a thymidine analog (see the formula below).

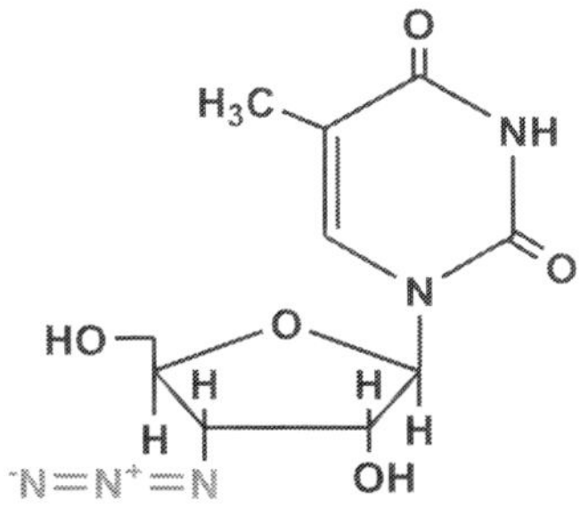

**Zidovudine**

The newly synthesized viral dscDNA enters the nucleus of the infected host cell and integrates itself into a host chromosome catalyzed by the *viral Integrase enzyme*. The viral genes are then transcribed into mRNAs to be translated into viral proteins and also into genomic RNA that combines with the viral proteins to form new viral particles; See Figure 42.

> *Worth Noting: Reverse Transcriptase applications*
>
> Reverse Transcriptase is a very important tool for recombinant DNA technologies mainly for gene cloning at mRNA level through conversion of mRNA into dscDNA. This is because mRNA is the smallest net structure of any gene that could be manipulated in vitro; however, it is very unstable in vitro to be utilized in recombinant DNA technology, that necessitates its conversion into the stable cDNA. It is also used for studying level of gene expression at mRNA level using the polymerase chain reaction.

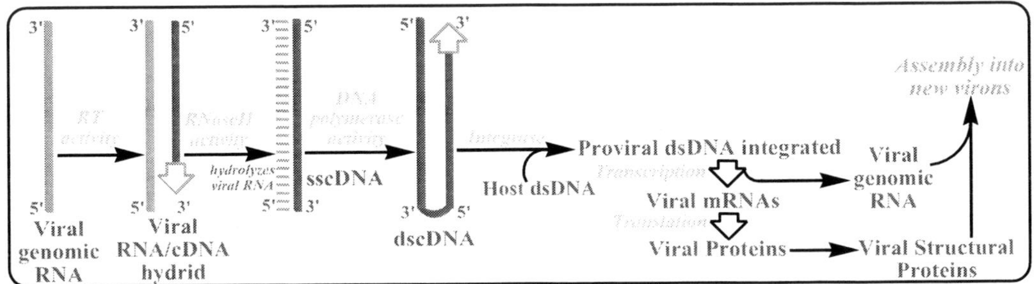


Figure 42. The role of reverse transcriptase in the life cycle of retroviruses.

The presence of pseudogenes in eukaryotes with net gene structure complementary to a mature mRNA sequence, e.g., those of globins and immunoglobulins, may suggest presence of reverse transcriptase activity in human cells, too.

## The Gene Concept and Structure

A gene is the smallest physical and functional unit of inheritance encoded in the DNA sequence of an organism (chromosomal or plasmid) that codes for a specific functional RNA (mRNA, tRNA, rRNA, snRNA, etc). It is an integrated functional unit composed of the *gene proper* (that is transcribed into RNA and mostly contains non-expressed intervening *intron* sequences and expressed *exon* sequences), and the transcription *regulatory sequences* (*response elements*, *promoter* and *terminator*). Genes vary massively in size ranging from 3000 bp for the shortest to 2.4 million bp for the largest known human gene, the dystrophin. The mRNA codes for a polypeptide chain or even a part of a polypeptide, whereas, tRNA and rRNA execute their functions as they are in protein synthesis.

Each gene occupies a well-defined position on a chromosome, i.e., gene locus. Alleles of a gene are pair or series of related genes occupying the same locus on homologous chromosomes. This phenomenon is called gene polymorphism that could be native and functionally normal, or, acquired by diseases-related mutations.

These alleles may be similar (homozygous) or may be different (heterozygous). The gene action determines the genotype, i.e., whether these alleles are recessive, additive, cooperative, or dominant. In dominant inheritance one allele is dominant (i.e., expresses its effect on the phenotype) and the other allele is recessive (i.e., defective and non-functional and have no participation in the phenotype). The mercy of our Creator made the manner of inheritance of genes controlling most of the inherited disorders a recessive manner. This means that the two or more alleles of a gene need to be all defective before a disease appears. And, those causing deleterious diseases, e.g., cancer, have weak genetic penetrance that necessitates an additive/cooperative action of several genes.

In males, most of the genes carried on the sex X chromosome have no copies (alleles) on the other male sex Y chromosome. Therefore, in sex-linked inherited diseases, females are usually carriers of the diseases, whereas, males are more susceptible because of having one allele.

*Worth Noting: The genetic insulators*
The *genetic insulators* are 1-2 kb sequences that bind specific protein(s) and flank and delimit functionally coordinately regulated number of genes in a genetic domain. They abolish the chromatin positional effect, e.g., euo- vs. hetero-chromatin, so as the insulated domain would be expressed in either case by inducing specific chromatin modifications. They also prevent 'cross-talk' with regulatory sequences from adjacent genetic domains. *The locus control region* is a larger modified form of the insulator that has same properties along within being regulatory for the developmental stage- and tissue-specific expression of the genetic domain. Mutations disrupting these insulating/regulatory sequences are implicated in some inherited diseases, e.g., some forms of β-thalassemia.

## The Structure of the Gene

The gene is a transcription unit DNA sequence composed of upstream regulatory sequence(s), promoter sequence(s), the coding gene proper sequence, termination sequence(s), and downstream regulatory sequence(s). Whenever presented, the sequence of the regulatory sequences; promoter, response elements and insulators is always described in the 5'-3' direction in the coding strand:

- *The gene response elements (enhancer and silencer)* are *cis-acting* response sequences that regulate the above-basal rate of gene expression. They bind specific proteins (*transacting factors*) that in turn interact with the basal transcription machinery to facilitate or prevent its DNA binding and activity.
- *The gene promoter region and the transcription initiation start point* that mostly locates up-stream the gene proper. The promoter provides the region of binding of the RNA polymerase and control the basal rate of transcription.
- *Transcribed region or gene proper* is the DNA sequence that is copied as hnRNA or other types of RNAs that is mostly composed of introns and exons in eukaryotes, but introns-containing genes are very rare in prokaryotes.
- *Termination region*, a regulatory DNA sequence down-stream the gene proper of some genes, at which the RNA polymerase disassembles from the DNA template. A gene may have several terminator sequences that are differentially use to suites specific tissue needs.

## Characteristics of the Gene Promoter

Promoter is a modulatory DNA sequence locates around a transcription initiation site – and mostly upstream the gene proper. It is composed of a complex array of cis-acting regulatory elements required for efficient and accurate initiation of transcription and for controlling expression of a gene. It is the transcription initiator sequence and works in the same orientation of the gene reading frame (3'⇒5') and is located in a close continuity with the gene proper. RNA polymerase binds DNA template at the gene promoter to initiate the basal (unregulated) rate of transcription.

Some promoters are weak and others are strong with much faster rate of transcription; nevertheless, there are certain promoterless genes. One gene may also have more than one promoter to be used differentially according to tissue specific needs.

The promoter sequence, its minimum requirements, conserved regions and regulation are identified through several strategies that include:

- Site-directed mutagenesis (chemically or using mismatch or degenerate primers) used to generate 5'-deletions of upstream DNA sequences and or missense mutations to determine regions essential for transcription initiation
- Clone into a vector carrying a reporter gene so as the promoter will be upstream the gene and the reporter gene expression is analyzed, e.g., β-galactosidase in E. coli
- Likewise, the reporter gene, e.g., luciferase or Syber green proteins could be transfected into eukaryotic cultured cells so as to be studies in eukaryotic cell milieu.

*Characteristics of an example prokaryotic promoter from the E coli:* it is simpler than eukaryotic promoter and is formed of three parts (see Figure 43):

- *Pribnow or TATA box* is a 6 bp sequence; 5'-TATAAT-3' located 10 nucleotides upstream (i.e., to the left of) the transcription start site +1 (i.e., the first transcribed nucleotide).
- *The two spacer stretches*; one is ~10 bp locates between the transcription start and Pribnow box and the other is ~19 bp locates between the Pribnow box and the -35 region
- *The -35 region* is a stretch of 8 bp; 5'-TGTTGACA-3' locates about 35 nucleotides upstream the transcription start site +1. Both Pribnow box and the -35 region are recognized by RNA polymerases.

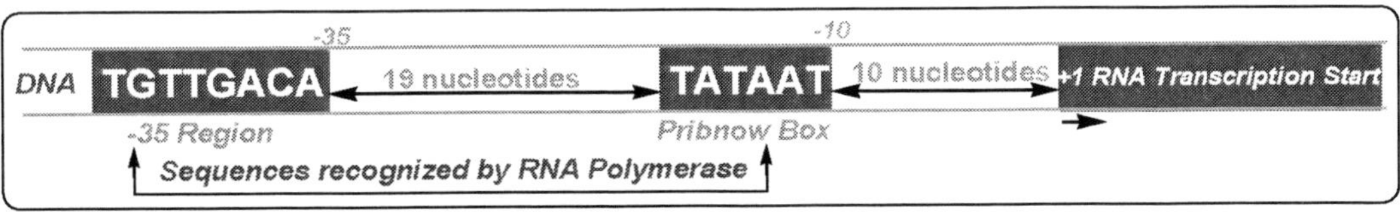


Figure 43. Example prokaryotic promoter structure.

*Characteristics of an example eukaryotic Promoter:* the eukaryotic promoter region is more complex than the prokaryotic one and consists of at least three parts (see Figure 44):

- *Hogness or TATA box:* It is an 8 bp stretch of DNA, e.g., 5'-TATAAAAG-3' located 25 - 30 nucleotides upstream the transcription start site +1. ~30% of the eukaryotic genes do not have TATA box and instead they have *initiator sequence* that spans the transcription start site –3 to +5, e.g., TCA[G/T]TT/C. ~30% of the stronger genes have both TATA and initiator sequences. Still ~25% of the genes have initiator sequence and a *downstream promoter sequence* ([A/G]G[A/T]CGTG, 25 bp downstream (i.e., to the right of) transcription start site +1. The remaining ~15 of the genes contain the three elements. These elements determine where transcription starts through the binding of the TATA binding protein (TBP) and its associating factors (TAFs) to form the TFIID complex that recruits the RNA polymerase

- *The spacer stretches* of about 25 bp DNA between the transcription start and TATA box and about 40 nucleotides between the CAAT box and TATA box
- *Cis-acting upstream promoter elements*, e.g., *CAAT box*, a 9 bp stretch of DNA: 5'-GGNCAATCT-3' located about 70 to 80 nucleotide up-streams the site for start of transcription, and, the GC boxes, e.g., 5'-GGGCGG-3'. Specific trans-activating proteins bind these elements by their DNA-binding domain and interact with basal transcription machinery through their trans-activation domains. They control the frequency of transcription initiation for the basal rate of expression. Depending on their type, they are less rigid in their orientation dependency than the TATA box group. Many protein encoding genes do not have TATA boxes or initiator complexes and transcription can begin at multiple start sites over a 20-200 bp region. This gives rise to mRNAs of different lengths due to different 5'-ends. These genes are generally transcribed at low rates, e.g:, genes encoding enzymes of intermediary metabolism. Instead they have CG-rich regions of DNA 20-50 bp in length that are found ~100 bp upstream of the start site. Since CG-rich regions are very rare in vertebrate DNA, presence of CpG islands in genomic DNA suggests a transcription -initiation region. These CG islands are associated with about 20,000 genes in mammals.

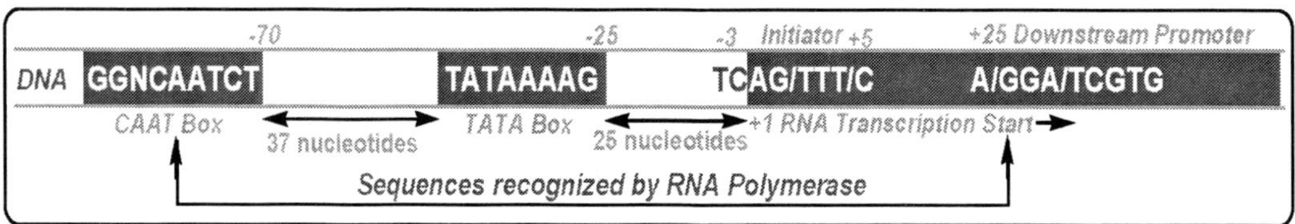


Figure 44. Example eukaryotic promoter structure.

Site-directed mutagenesis studies showed that single base change or deletion in the conserved boxes of the promoter dramatically decreases tare of RNA transcription as indicated by reporter gene expression, e.g., β-galactosidase. Mutations in the promoter sequence of the β-globin gene are the causative factors for several subtypes of β-thalassemia. These mutations reduces rate of expression of the gene and amount of the protein produced.

## Alternative Promoters

They are different promoter sequences that locate to different location relative to the transcription start point of a specific gene, e.g., Promoter 1 - Exon 1 - Promoter 2 - Exon 2 - Exon 3 - Exon 4. They could create mRNA having Exon 1 - Exon 2 - Exon 3 - Exon 4, or, Exon 1 - Exon 3 - Exon 4, or, Exon 2 - Exon 3 - Exon 4. They differentially regulate the rate of gene expression; quantitatively (absolute amount of mRNA), qualitatively (generation of structural isoforms of the gene mRNA through alternative exon usage that may encode diverse protein isoforms), tissue- and developmental stage-specifically, and differential mRNA stability and translation efficiency due to creation of diverse isoforms at the 5'-untranslated regions. Accordingly, they react differently to regulatory signals. However, they use an identical reading frame and produce identical protein sequences with exceptional cases of variation at the N-terminus due to leader exon escaping. Their presence leads to significant variation and complexity in the control of transcription rate and nature of mRNA generated. Alternative usage of such promoters is implicated in several diseases, e.g., cancers.

## Characteristics of the Gene Response Elements (Enhancer/ Silencer)

They are transcription regulatory sequences that may be remote from the gene by hundreds or thousands of nucleotides up- or down-stream the gene proper or even within the transcription unit. Therefore, it works in an orientation- and distance-independent manner. There are hundreds of types of response elements with rigid or flexible sequence. The response elements that activate transcription above the basal rate are called *Enhancer elements* and those that suppress transcription below the basal rate are called *Silencer/Repressor elements*.

They mediate responses to various signals including hormones, environmental changes and toxins such as dioxin though interaction with a large number (~ 2000 in human genome) of specific *transcription/transacting factors*. The best examples for these elements are the hormone-response elements for steroid-thyroid superfamily of hormones including vitamins D and A. In *testicular feminization* disease the patient is XY male but looks like a female with undeveloped male external genitalia (atrophied testes present in the inguinal canal). The patient synthesizes testosterone but has no intracellular testosterone receptors - that is a ligand-dependent transcription factor.

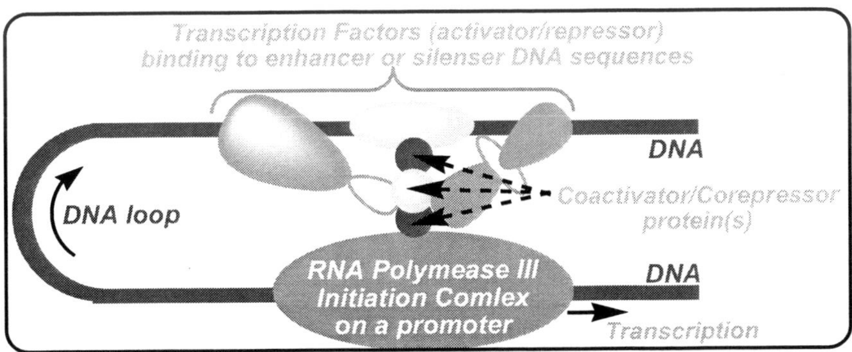


Figure 45. The basal transcription complex interacts with transcription factors through coactivators and corepressor that bind their target DNA sequences at the major groove.

*Transcription factors* are activator or suppressor proteins that bind at these elements and interact with RNA polymerases directly or indirectly through coactivator or corepressor proteins to increase or decrease rate of transcription; see Figure 45. Transcription factors have certain specific kind of secondary structures, i.e., motifs for: specific DNA binding (a helical DNA binding domain), dimerization domain (if the transacting factor binds as a dimer), specific ligand binding domain, and transacting domain that interacts with the basal transcription machinery. Several Zinc-fingers, leucine zipper, helix-turn-helix and helix-loop-helix per protein are common secondary motif structures in the DNA-binding domains that enable high affinity/site-specific DNA binding as the result of complementarity between the surface of the protein domain and specific region of the double stranded DNA sequences.

Like other DNA-binding proteins and enzymes, they bind through secondary bonds, e.g., hydrogen and ionic bonds with the DNA sugar-phosphate backbone. However, the specificity of their DNA binding come from the proteins making multiple contacts to the edges of the DNA bases, allowing them to "read" the specific DNA sequence – also through secondary bonds. Most of these base-interactions are made in the major groove, where the bases are most accessible. Transcription factors also modulate rate of gene expression at the transcription level through binding and helping the RNA polymerase, either directly or through other mediator proteins to locates at the promoter (e.g., σ actor) and allows it to begin transcription. Alternatively, transcription factors can bind and modulate the activity of enzymes that modify the histones (e.g., acetylases/deacetylase) at the promoter region. This will change the accessibility of the DNA template to the RNA polymerase for transcription. Mediator proteins (coactivators/corepressors) not only connect the regulatory transcription factor to the initiation machinery but they also could converge effects of several regulatory transcription factors on several regulatory sequences to affect assembly and progression of one initiation complex on one promoter. The multi-nucleoprotein complexes formed on one eukaryotic promoter maybe contain as much as 100 polypeptides with a few megadaltons mass.

These domains are composed of α-helices, β-sheet and turns. Zinc fingers are small protein domains that can coordinate one or more zinc ions to help stabilize their folds. Zinc fingers coordinate zinc ions with a combination of cysteine and histidine residues. They can be classified by the type and number of these zinc coordinating residues, i.e., Cys2His2, Cys4, and Cys6. Typical examples are the nuclear hormone receptors. Various strategies have been developed to engineer Cys2His2 zinc fingers to bind desired sequences. Such engineered zinc finger arrays can then be used in numerous applications such as artificial transcription factors, zinc finger methylases, zinc finger recombinases, and Zinc finger nucleases. Artificial transcription factors with engineered zinc finger arrays have been used in numerous scientific studies and an artificial transcription factor that activates expression of vascular endothelial cell growth factor is currently being evaluated in humans as a potential treatment for several indications. An ongoing clinical trial is evaluating Zinc finger nucleases that disrupt the chemokine receptor CCR5 gene in CD4+ human T-cells as a potential treatment for HIV/AIDS. Cys4 constitutes the steroid-thyroid nuclear receptor group of transcription factors. Each half of a leucine zipper dimerization motif consists of a short alpha-helix with a leucine residue at every seventh position. The standard 3.6 residues per turn alpha-helix structure changes slightly to become a 3.5 residues per turn alpha-helix. Known also as the heptat repeat, one leucine comes in direct contact with another leucine on the other strand every second turn. The bZip family of transcription factors consists of a basic region which interacts with the major groove of a DNA molecule through hydrogen bonding, and a hydrophobic leucine zipper region which is responsible for dimerization. Leucine zipper regulatory proteins include c-fos and c-jun (the AP1 transcription factor), important regulators of normal development and cell cycle, and, the metabolic regulator cAMP-responsive-element-binding protein; CREB (activated after phosphorylation by PKA). If they are overproduced or mutated in a vital area, they may generate cancer. These proteins interact with the DNA as dimers (homo- or hetero-) and are also called basic zipper proteins (bZips); see Figure 46.

In humans, there are ~2000 proteins with transcription controlling activity. Mutational inactivation of repressor proteins can result in very high levels of protein expression with devastating consequences and disease implications. Likewise, inactivating an activating transcription factor or its abnormal upregulation is also having devastating fate. As the most dynamically active part of the proteome of a cell, the level of transcription factors expressed and activated in a particular cell vary at a particular time and metabolic state.

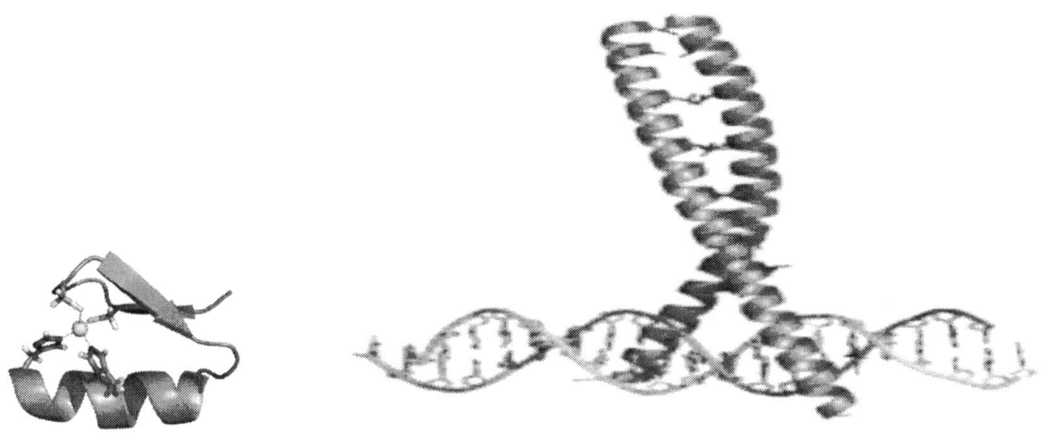

Figure 46. The Cys2His2 zinc finger motif (to the Left), consisting of an α helix and an antiparallel β sheet. The zinc ion (green) is coordinated by two histidine residues and two cysteine residues. Leucine Zipper (blue) bound to DNA (To the right). The leucine residues that represent the 'teeth' of the zipper are colored red.

## *Signal Transduction of the steroid/Thyroid Intracellular Nuclear Receptor Superfamily as Ligand-transcription Factors*

*The group I water-insoluble lipophilic hormones* (steroid-thyroid superfamily/ calcitriol/retinoic acid/derived fatty acid/eicosanoids) are able to cross the cell membrane. Therefore, their receptor is located to the cell cytoplasm or even the nucleus (in case of the thyroid hormones). Specific binding of the hormone and its receptor produces the functional hormone-receptor complex that is a ligand-dependent transcription regulatory factor. The later attains a conformation and phosphorylation change that enables its nuclear translocation and DNA binding. The receptor-hormone complex is the second messenger for group I hormones - that directly controls the rate of specific gene expression through binding to specific response DNA sequences.

The steroid hormone/thyroid hormone superfamily of receptors is structurally similar and resides primarily in the nucleus (all except glucocorticoids), or cytoplasm (glucocorticoids). They are single polypeptide with 6 structural domains; A-F. The A/B at the N-terminus is the transcription activating function 1 (AF1) region, the C domain is the DNA-binding domain, the D domain is the flexible hinge between C and the ligand-binding E domain that also contains at its C-terminus the second transcription activating function 2 (AF2), and the F region is poorly defined functionally (Figure 47). The nuclear localization signaling sequences locate to the DEF domains. For this superfamily of signalling effectors, the primary messenger is the hormone and their second messenger is the intracellular functional hormone-receptor complex that functions as a transcription regulatory function for its target gene(s).

The dimeric ligand-bound receptor binds specifically to its DNA response element. The element sequence is a similar partially palindromic inverted repeat; ⇐5'-GGTACA-NNN-TGTTCT-3'⇒ for a homodimeric glucocorticoids (GRE), progestins (PRE), mineralocorticoids (aldosterone; MRE) and androgens (testosterone and dihydrotestosterone; ARE) receptor binding; where N is any nucleotide. The binding specificity is conferred through the differential; specific ligand/receptor availability, flanking DNA sequences, accessory elements and/or the coactivators/corepressors used. The DNA response element for estrogens (estrone, 17β-estradiol and estriol; ERE) is also an inverted repeat; ⇐AGGTCA---TGA/TCCT⇒ for the homodimeric receptor binding. The thyroid hormones (triiodothyronine and thyroxine; TRE), retinoic acid (RARE), and vitamin D (calcitriol; VDRE) DNA response element is similar to the estrogen one but in a direct repeat form; ⇒5'-AGGTCA-$N_n$-AGGTCA-3'⇒ with a spacer number of nucleotides that is specific for each type; where, n = 3 for TRE, =4 for RARE and = 5 for VDRE. Unlike the other steroids, the receptors of these hormones bind as a heterodimer with the 9-cis-retinoic acid receptor, RXR.

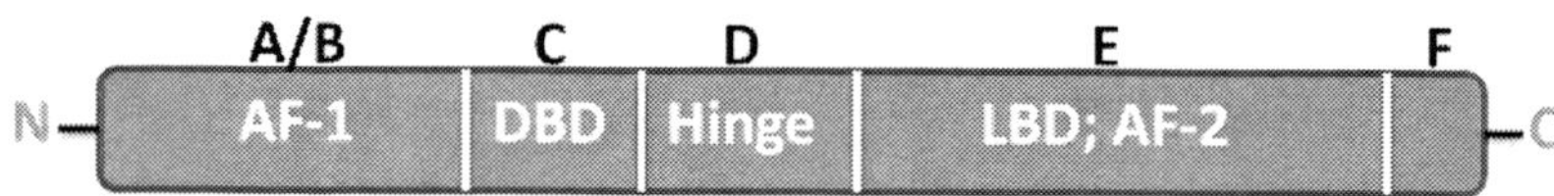


Figure 47. The structural domains of the nuclear receptor superfamily. AF-1 is the transcription activating function 1, DBD is the DNA binding domain, LBD is the ligand binding domain, and, AF-2 is the transcription activating function 2.

Similar to the VDR, TR, and RAR, the nuclear receptors; peroxisome proliferator-activated receptors (PPARα, β and γ), the liver X-activated receptor (LXR), the farnesoid X-activated receptors (FXR), xenobiotic X receptor (CAR) and the pregnane X receptor (PXR) form heterodimers with the 9-cis-retinoic acid specific RXR and bind to their appropriate response elements in DNA in an inactive state. Binding of their respective specific ligand (oxysterols for LXR, farnesol and bile salts for FXR; androstanes, phenobarbital and xenobiotics for CAR; pregnanes, xenobiotics and secondary bile salts for PXR, and fatty acids and their derivatives, eicosanoids, and the hypoglycemic thiazolidinediones for the PPARs) activates the complex to control rate of gene expression. PPARα and β are implicated in the proliferation of peroxisome and PPARγ is implicated in the adipocyte differentiation, and metabolism of lipid and carbohydrate. FXR is implicated in bile acid metabolism. LXR is implicated in cholesterol metabolism. PXR and CAR is implicated in protection against toxic metabolites, xenobiotics and certain drugs. These receptors regulate a variety of genes encoding cytochrome p450s (CYP), cytosolic binding proteins, and ATP-binding cassette (ABC) transporters to influence metabolism and protect cells against drugs and noxious agents. Their specific sequence direct repeat HRE binds the heterodimer with RXR and vary in the length of the pacer between the two half repeat (n = 1 for PPARs, = 3 for PXR, = 4 for FXR and LXR, and, = 5 for CAR).

The cytoplasmic glucocorticoid receptor is a multimeric complex associated with heat shock proteins. Cortisol (a glucocorticoid) binding induces receptor conformational change that releases the heat shock proteins and exposes the nuclear translocation signal for a dimeric form of the ligand-bound receptor. In the nucleus it binds the GRE. The thyroid hormone receptor exists in a DNA bound form to its response element. The hormone binding changes its activity through modulating nature of its protein interactions to modulate rate of gene expression.

*Tamoxifen* is a synthetic estrogen competitive antagonist that exploits the ligand specificity of the steroid-receptor interaction to bind the receptor and obliterate complex gene regulating ability to prevent proliferation of the estrogen-dependent breast cancer that require the continuous presence of the hormone estrogen around its cells. Likewise, synthetic progesterone competitive antagonists, e.g., *mifepristone*, are used to terminate early (preimplantation) pregnancies through binding progesterone receptor and blocks hormone actions essential to implantation of the fertilized ovum in the uterus. See the formulae of these compounds in Figure 48.

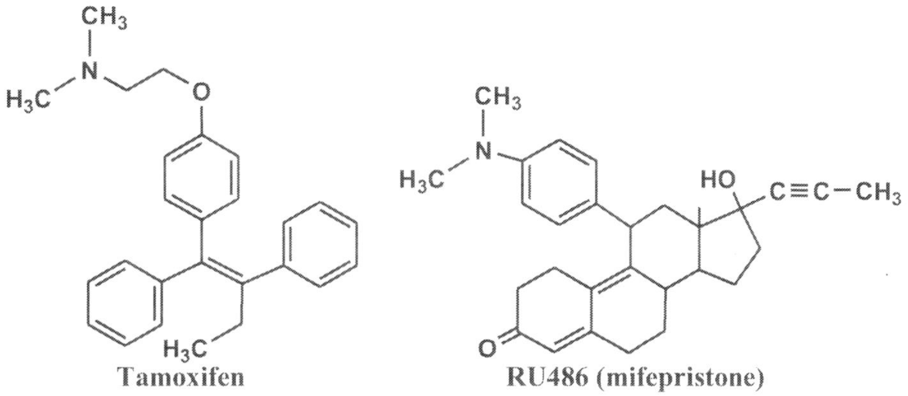

Tamoxifen

RU486 (mifepristone)

Figure 48. Formulae of estrogen (Tamoxifen) and progesterone (mifepristone) antagonists.

However, at least some steroids also act through plasma membrane receptors by a completely different mechanism. The classic mechanism for steroid hormone action through nuclear receptors does not explain certain effects of steroids that are too fast to be the result of altered protein synthesis. For example, the estrogen-mediated dilation of blood vessels is known to be independent of gene transcription or protein synthesis, as is the steroid-induced decrease in cellular cAMP concentration. Another transduction mechanism is probably responsible for some of these effects. A plasma membrane protein predicted to have seven transmembrane helical segments binds progesterone with very high affinity and mediates the inhibition of adenylate cyclase by that hormone, accounting for the decrease in cAMP level. A second non-classical mechanism involves the rapid activation of the Mitogen Activated Protein kinase (MAPK) cascade by progesterone, acting through the soluble progesterone receptor. This is the same receptor that, in the nucleus, causes the much slower changes in gene expression that constitute the classic mechanism of progesterone action. How the MAPK cascade is activated is not yet clear. Not only the group I hormones acting through ligand-dependent transcription factors but also membrane-bound receptor that signal for group II hormones act in part through modulation of intracellular cAMP that activates cAMP-dependent protein kinase that phosphorylates target transcription factors, e.g., CREB and activation transcription factor (ATF) that modulate the rate of gene expression through binding their respective DNA response sequences, e.g., the palindromic cAMP response element (CRE; TGACGTCA).

*Worth Noting: Several response elements regulate one gene: Phosphoenol-pyruvate carboxykinase gene as a model*

Tens of response elements with the specific transcription factors binding to them may work on one gene in a cooperative, agonistic, synergistic or antagonistic fashion. This brings about the tissue-specific pattern of gene expression such as albumin gene in the liver. In the liver, the expression of the *phosphoenol-pyruvate carboxykinase gene* – an enzyme key regulatory in gluconeogenesis – is under inhibitory effect of insulin through an insulin-response element and under stimulation by; glucagon through a cAMP-response element, thyroxine through thyroid hormone-response element and glucocorticoids through glucocorticoid-response element (see Figure 49). An activating transcription factor on one gene could be a repressor transcription factor on another gene. Their activity is under control through several mechanisms including; their gene transcription rate, phosphorylation, and availability of synergistic and/or antagonistic proteins. Some genes have none of such elements, e.g., the constitutional or housekeeping genes that have a constant rate of basal transcription.

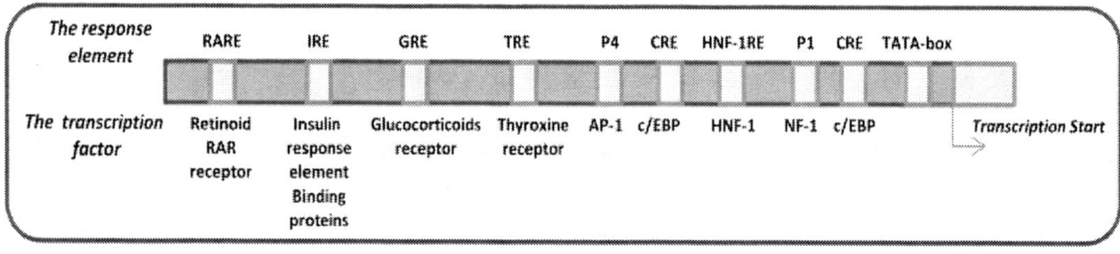


Figure 49. Gene response elements that regulate the expression of phosphoenol pyruvate carboxykinase. These elements are distributed over ~1 kbp. They include response element for each of; the retinoic acid (RARE), insulin-response element binding proteins (IRE), the glucocorticoid receptor (GRE), the thyroxin receptor (TRE), Activator Protein-1; AP-1 (P4), cAMP-activated protein kinase A-activated transcription factor C/EBP (CRE), hormone-independent hepatocyte nuclear factor-1; HNF-1 (HNF-1RE), and hormone-independent factor nuclear factor-1; NF-1 (P1). The effect of these proteins is integrated through coactivator/corepressor to enhance or suppress the basal transcription complex. There transcription regulators are stimulatory with the exception of IRE and its binding proteins that prevent the binding of the glucocorticoids receptor to GRE.

# Transcription (RNA Synthesis)

Transcription is the process of synthesizing a complementary RNA copies from specific regions along the length of gene DNA, i.e., the gene proper, using DNA-dependent RNA polymerase (Pol) and ribonucleotides and the base pairing role of A⇒U and C⇒G. The general characteristics of the RNA polymerases from eukaryotes and prokaryotes are compared in Table 5. These enzymes do not have proofreading ability. Like DNA polymerases, RNA polymerases have proof-reading characteristics and require help from topoisomerases. The mitochondrial RNA Pol is similar to the prokaryotic RNA polymerase.

**Table 5. Prokaryotic vs. eukaryotic DNA-dependent RNA Polymerases**

| In Prokaryotes | In Eukaryotes |
|---|---|
| A single type of RNA Pol synthesizes all types of RNA. | There are 4 types of RNA polymerases; each is specific for the synthesis of a specific type of RNA. |
| Products of RNA Pol require slight or no modification after transcription. | Mostly require extensive post-transcriptional modifications. |
| The holoenzyme ($\alpha_2\beta\beta'\sigma$) is formed of five subunits: two identical $\alpha$ subunits, two similar but not identical $\beta\beta'$ catalytic subunits and a regulatory sigma ($\sigma$) subunit. The core part of the enzyme is formed of the four $\alpha_2\beta\beta'$ subunits. The specific regulation of gene transcription comes from the different $\sigma$ factors specific to every gene. The several types of sigma factors are designated $\sigma^{54}$, $\sigma^{70}$, etc, with the later as the major type and functions in initiation, and, the former is essential for interaction of the basal machinery with activators and repressors. | They are much complex in structure and are formed of 2 large (homologous to $\beta\beta'$) and up to 14 small subunits. The three types of RNA polymerases are: *RNA Pol I*: Responsible for synthesis of the large RNA molecules (rRNA). It is insensitive to $\alpha$-amanitin (a peptide toxin from the toxic red mushroom. It prevents polymerase translocation. *RNA Pol II*: Responsible for synthesis of mRNA and some snRNA. It is sensitive to low concentration of $\alpha$-amanitin. *RNA Pol III*: Responsible for synthesis of small RNA molecules, i.e., tRNA, 5S rRNA and snRNA. It is sensitive to high concentration of $\alpha$-amanitin. *Mitochondrial RNA Pol*: Is similar to the prokaryotic RNA polymerase. |

## Steps of Transcription

### *i) Initiation*

Initiation of transcription occurs on a single strand of the transcription unit (a gene) that is called template (aka, non-coding or anti-sense) strand that would be complementary to the produced RNA, and it never occurs in the other strand, non-template (aka, coding or sense) strand that is similar to the RNA product except for the uracil/thymine replacement. RNA polymerases lack the 3'⇒5' exonuclease proofreading activity. RNA polymerases are zinc containing metalloenzymes. Several RNA polymerases may transcribe same gene simultaneously but in a phase and spaced manner.

Initiation is regulated and initiated by; chromatin decondensation by histone deacetylases, interaction of the basal transcription machinery with regulatory (positive and negative) signaling proteins that enhances or hinders the assembly of the machinery, and, by phosphorylation on serine/threonine residues of the heptapeptide repeats (YSPTSPS) of C-terminal domain of the polymerase II. Phosphorylation is necessary for initiation (serine 5) and further phosphorylation enables the shift into elongation (serine 2). Also, specific drugs inhibit initiation such as riphampicin.

*TATA box Binding Protein (TBP)* binds to the TATA box sequence to induce a conformational change in the DNA that enables *the general transcription factors* (TFs; ~ 20 different subunits) to interact with the promoter area. They stabilize the binding of TBP to the promoter region and help selecting the transcription start site and recruits RNA Pol II. The complex formed is called the preinitiation complex with the C-terminal domain repeats of RNA Pol II unphosphorylated. One of these transcription factors (TFIIH) has helicase activity to unwind the DNA duplex at the transcription start site requiring hydrolysis of ATP. Transcription starts in situ and moves ahead, the C-terminal domain repeats of RNA Pol II get serine/threonine phosphorylated to release the transcription factors. In comparison RNA polymerases I and III require different polymerase-specific general transcription factors, recognize different regulatory sequences, and they do not get phosphorylated to progress nor require ATP hydrolysis.

Mutations in any subunit of the basal transcription machinery that disable it from interacting with the general or specific coactivators/corepressors and hence to regulatory activator and repressor proteins, or, mutations in such regulatory proteins that disable them from interacting with the basal transcription machinery could lead to serious diseases. Examples include the dominant mutation (i.e., one mutant and one normal gene copy) in gene encoding CBP/p300 (CREB-binding protein) coactivator (into which converge several transcriptional regulatory signals) causes severe developmental disorder, e.g., Rubinstein-Taybi disease, and mutant transcriptional activator Huntington's protein is unable to interact with CBP to enables its binding. As a consequence, the transcription machinery is repressed and the expression of key genes required for striatal neuronal survival is prevented resulting in Huntington's chorea with a late onset neurodegenerative disorder due to death of these cells.

Like DNA replication, the template strand is read in 3'⇒5' so that the synthesis of the RNA goes in 5'⇒3' but it does not require a primer because RNA Pol has an initiation and an elongation nucleotide binding sites. Which DNA strand is the gene template and which is the coding, differs from one gene to the other but is always the strand that contains an *open reading frame* in 3'⇒5' direction which can be used by the RNA Pol as a template to synthesize an RNA; see Figure 50. Therefore, an open reading frame is a DNA codon sequence - with a start of transcription codon and a termination sequence - transcribable into RNA; i.e., if it is an mRNA, it would have a suitable size, a start of translation and a stop of translation codons. The inhibition of eukaryotic polymerase II by α-amanitin after accidental eating the red poisonous mushroom - Amanita phalloides - causes gastrointestinal disturbances, electrolyte loss, fever and the most patients die due to liver and kidney dysfunction.

In prokaryotes, RNA Pol recognizes the promoter region by the help of the σ (sigma) factor that then recruits the core enzyme ($\alpha_2\beta\beta'$) that tightly binds DNA. There are several types of σ factors that recognize different promoters to differentially control gene(s) transcription according to the growth and environmental conditions of the bacterium. Interaction among the polymerase, sigma factors, and activators and repressor proteins determines rate of specific gene expression. Examples of these sigma factors include:

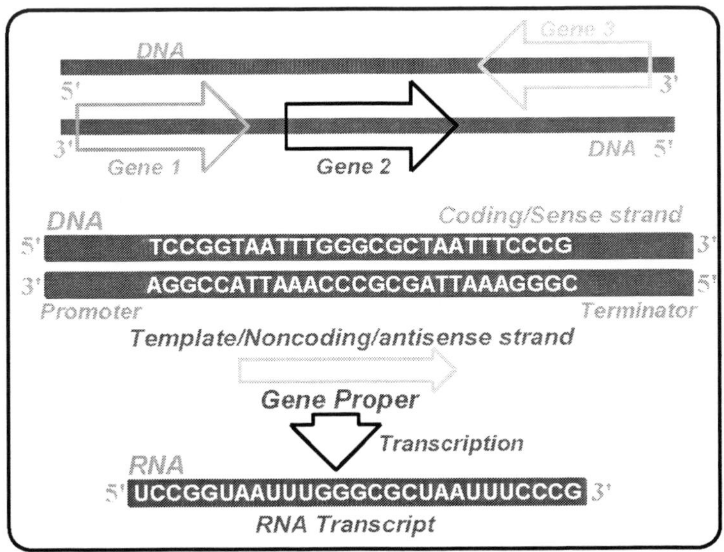


Figure 50. The template/noncoding/antisense strand is used for transcription (on the left) and gene template location may be on any of the two strands and the direction of transcription differs dependently (on the right).

- $\sigma^{70}$ (RpoD) - the "housekeeping" sigma factor, controls most genes in growing cells
- $\sigma^{54}$ (RpoN) - the nitrogen-limitation sigma factor
- $\sigma^{38}$ (RpoS) - the starvation/stationary phase sigma factor
- $\sigma^{32}$ (RpoH) - the heat shock sigma factor, for the cell to face heat shock
- $\sigma^{28}$ (RpoF) - the flagellar sigma factor
- $\sigma^{24}$ (RpoE) - the extracytoplasmic/extreme heat stress sigma factor
- $\sigma^{19}$ (FecI) - the ferric citrate sigma factor, regulates the fec gene for iron transport.

The core enzyme unwinds 17-20 bp nucleotides to separate the two strands (~60 bp in eukaryotes, from –30 to +30), i.e., the transcription bubble. This requires unwindase, disruption of the nucleosome structure and the participation of topoisomerases. The enzyme complex then searches for the transcription initiation site +1 (an open reading frame starting at TAC).

The synthesized RNA always starts with a purine that enters at *the initiation site* of the enzyme – that is mostly A with exceptional C in genes without a TATA box. This 5'-purine ribonucleotide at the 5'-end retains its triphosphate and stays in mature mRNA.

When the second nucleotide enters at *the elongation site* of the β-unit of the enzyme it forms a phosphodiester bond with the 3'-OH group of ribose of the first nucleotide and its phosphate-5'-OH on C5 of the ribose with the release of a pyrophosphate (PPi). While on the promoter, the polymerase synthesizes 10 – 20 nucleotide stretch of RNA.

### *ii) Elongation*

The σ Factor is released before elongation starts leading to core enzyme conformational change that allows its translocation to clear the promoter. The four ribonucleotides triphosphate; ATP, GTP, CTP and UTP continue to enter into the polymerization (or elongation) site with the release of a pyrophosphate (PPi) each time a new nucleotide is added to the growing RNA chain. Pyrophosphatases hydrolyze PPi into 2 Pi to ensure irreversibility of the polymerization. The RNA Pol continues transcription from 3' towards 5'-end of the template strand according to the base pairing role in an anti-parallel manner. Elongation continues through the termination sequence.

Elongation is regulated by drug (e.g., actinomycin D, and streptoglydigin) and sequence-dependent arresting factors and RNA Pol II catalysis-improving factors that ensure fine toning of the rate of elongation. Elongation is also down-regulated by blocking polymerase progress and/or by deactivating the polymerase. Chromatin structure-oriented factors through histone modification (phosphorylation, acetylation, methylation ,and ubiquination) interacting with the polymerase also control its progression rate.

### *iii) Termination*

At the completion of the transcription of a gene, the elongation has to be terminated. The termination takes place in two ways viz. *a) rho (ρ) factor dependent*, or, *b) rho (ρ) factor independent:*

a. *Rho-dependent termination:* The rho (ρ) factor is an ATP-dependent RNA-stimulated helicase that recognizes and binds the termination sequence (~40 bp) in the template DNA to disassemble the enzyme/RNA/DNA complex.
b. *Rho-independent termination:* RNA polymerase stops transcription when the synthesized complete RNA molecule takes its three-dimensional form with the *formation of specific secondary structures such as a hairpin followed by an oligo-U sequence.* This structure signal to pause the RNA polymerase. The inherent character of paused RNA polymerase prevents further polymerization and signal disassembly of the transcription machinery.

The Pol II termination sequence may occur hundreds of base-pairs (0.5 - 2 kb) downstream the site at which poly(A) tail is added in mRNA. An RNA endonuclease cleaves the primary RNA transcript at 11 - 31 bases 3' of the *cleavage/polyadenylation consensus sequence, i.e., AAUAAA*, where, the polyA tail is added to the new 3'-end. Eukaryotic termination may be more complex and involves the polyadenylate tail addition signal and a destabilizing AU sequence. One gene may have more than one termination sequence and tissues would use them differentially according to their needs. Figure 51 summarizes steps of transcription.

*RNA polymerase III (Pol III)* transcribes DNA to synthesize the housekeeping ribosomal 5S rRNA, tRNA and other small RNAs. The regulation of the rate of Pol III transcription requires fewer regulatory proteins than RNA polymerase II. Compared to Pol II, Pol III normally relying on internal control sequences within the transcribed gene proper and does not require control sequences upstream of the gene. However, upstream sequences are occasionally used as in the case of U6 snRNA gene with an upstream TATA box. Polymerase III terminates transcription at small polyTs stretch. Unlike PolII, most of the components of the basal transcription machinery remain bound to the Pol III that ensures a high rate of transcriptional reinitiation of the target genes. Unlike Pol II, initiation by Pol I 1nd III does not require ATP hydrolysis and there are no cap or poly(A) tail added to their products.

*RNA polymerase I (Pol I)* is, in eukaryotes, the only enzyme that transcribes ribosomal RNA (except 5S rRNA) in the nucleolus into one large transcript that is subsequently cleaved by snoRNA into 18S, 5.8S and 28S rRNA molecules of the ribosome in eukaryotes. Because of the simplicity of Pol I transcription it is the fastest acting polymerase. In cells active in protein synthesis, high rate of translation is coordinated to a high rate of rRNA synthesis through; determining the number of rRNA genes transcribed at a specific time among the several hundreds of genes available, and, modulation of the rate of transcription without changing the number of actively transcribed genes. The less differentiated cells have a higher rRNA synthesis and a larger number of rRNA genes transcribed. When rRNA synthesis is stimulated, SL1 (selectivity factor 1) binds the promoters of rRNA genes that were previously silent and recruit a pre-initiation complex to which Pol I will bind and start transcription of rRNA. Instead of having TATA box in the promoter, Pol I has UCS (Upstream Control Sequence) and its binding factor (UCF) that recruit the basal transcription machinery. Elongation is likely to be interrupted at sites of DNA damage. Transcription-coupled repair occurs similarly to Pol II-transcribed genes and require the presence of several DNA repair proteins, such as TFIIH, CSB and XPG. Once the promoter is cleared off Pol I, UBF and SL1 remain promoter-bound to recruit another Pol I to ensure transcription of rRNA genes multiple times simultaneously – unlike Pol II that associate its gene in one-to-one ratio at a time. Pol I requires removal of DNA supercoiling by topoisomerases. Specific transcription factors and an T-rich region help terminating its transcription.

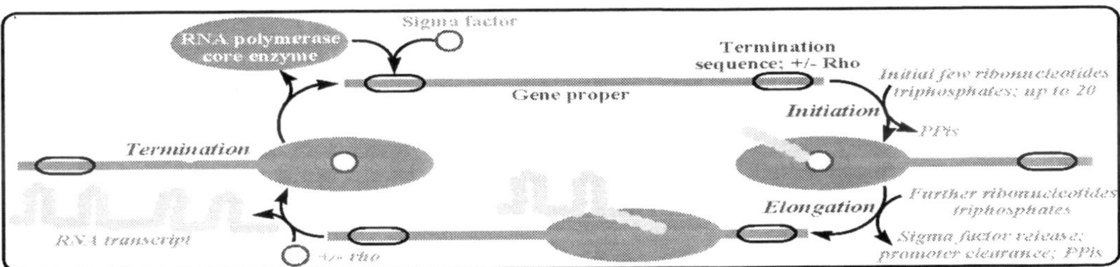


Figure 51. The three major transcription steps illustrating the role of the sigma and rho factors.

## *Regulation and Antibiotic Inhibitors of Transcription*

- *Covalent modification:* Eukaryotic RNA polymerase and the other transcription regulators are under control by activating *phosphorylation-inactivating dephosphorylation mechanisms*
- *Rifamycin* binds to the core enzyme and occupies the substrate binding site to prevent entrance the incoming nucleotides to the initiation site of the prokaryotic system. It also affects the mitochondrial transcriptase albeit at a massively higher concentration than required to kill bacteria
- *Actinomycin D* and *daunomycin* intercalate the DNA helix and inhibits its transcription by preventing translocation of RNA polymerase along DNA
- *Streptoglydigin* binds the β-subunit of prokaryotic polymerase and inhibits the transcription elongation
- *Heparin* is a polyanion that binds to the β'-subunit of the RNA polymerase and inhibits transcription *in vitro*.

# Post-transcriptional Processing of RNA

Processing of freshly-synthesized RNA transcript in the nucleus differs depending upon the type of RNA. The following are the specific posttranscriptional processing of mRNA, tRNA and rRNA. The initial transcript, thus, undergoes various modifications that create not only the salient functional characteristics of each type of RNA, but they also control the half-life and rate of usage of the mature RNA.

## Post-Transcriptional Processing of mRNA

Eukaryotic crude transcript of mRNA produced in the nucleus is called pre-mRNA or *heterogeneous nuclear RNA (hnRNA)*, i.e., the mRNA blue script. The newly synthesized hnRNA might be a complex transcript that carries information for several RNAs, i.e., polycistronic mRNA as in most prokaryotic mRNAs. Furthermore, initial mRNA transcript could contain the intervening sequences known as 'Introns' which have to be removed, whereas, the remaining 'Exons' are to be rejoined together, i.e., splicing as in most of the eukaryotic mRNAs.

Post-transcriptional processing of mRNA includes; 5'-capping; 3'-polyadenylation tailing, splicing (intron removal and alternative splicing of exons that also decreases the mRNA size), and, the post-maturation mRNA chemical modification by editing.

- *Capping:* It is the prompt addition of GTP to the 5'-end nucleotide-triphosphate of hnRNA activated by *guanylyl transferase* as a part of the cap-synthesizing complex bound to the phosphorylated C-terminal domain of the RNA Pol II at the end of the initiation phase of its transcription once the 5'-25 - 50 nucleotides are synthesized. *The cap is attached by 5'-triphosphate-5' linkage* with the loss of one phosphate ($\gamma$) from the original 5'-end nucleotide-triphosphate by a phosphatase and two phosphates from GTP during its addition. Subsequently, *guanine methyltransferase* adds a methyl group from S-adenosylmethionine to N7 of the guanine ring. C2 of the sugar in the first two original nucleotides and $NH_2$ on C6 of adenine of the first original nucleotide of the hnRNA are also subsequently methylated; see Figure 52. Among the functions of capping are; it enhances subsequent hnRNA processing; helps mRNA transport to cytoplasm; and controls rate of the translation and half-life of mRNA in the cytoplasm through specific binding to cap-binding proteins, e.g., eIF4A (see translation later). It also protects mRNA from the action of 5'⇒3' exonucleases and phosphatases, and therefore, controls mRNA half-life. RNA products of Pol I and III are not capped since the capping enzyme cannot interact with them and no need for the cap since they are untranslatable.

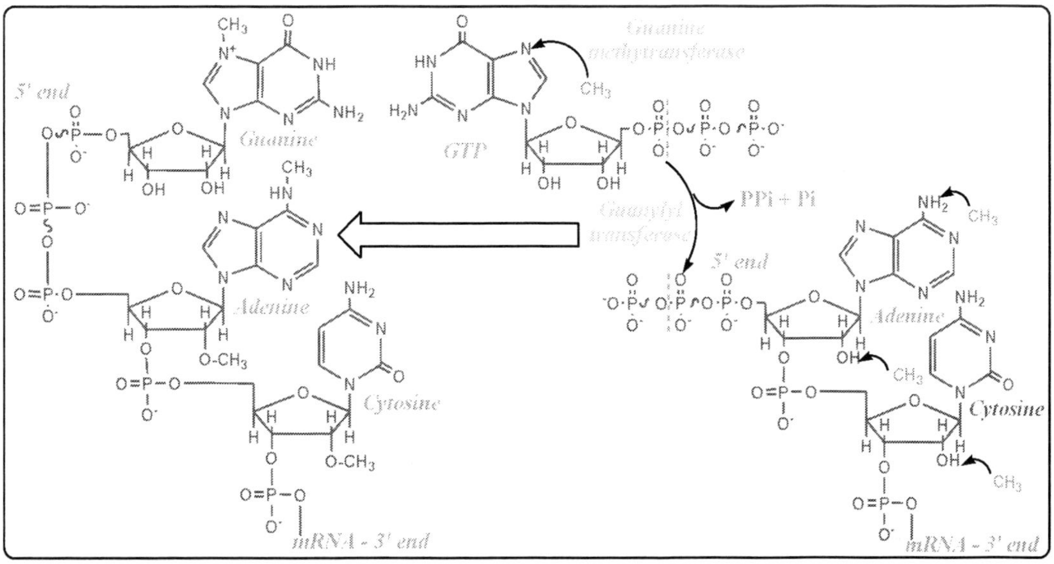

Figure 52. Addition of the 5'-triphosphate-5' GTP cap and its methylation along with methylation of the terminal original mRNA nucleotides.

- *Addition of polyadenylate 3'-tailing:* 20 - 250 ATP (polyadenylate = polyA tail) are added at the newly created 3'-end of mRNA as an integral part of the transcription termination process catalyzed by *polyadenylate polymerase*. With the help of three more proteins, the enzyme first recognizes the specific polyA addition signal, 5'-AAUAAA-3' within the mRNA 3'-UTR followed by another signal sequence, usually (5'-CA-3') which is the site of cleavage by the endonuclease activity of *polyadenylate polymerase* that cuts extra sequences at 10 - 35 bases 3'-downstream the 5'-AAUAAA-3' signal. As the poly(A) tails is synthesized, it binds multiple copies of poly(A) binding protein, which protects the 3'-end from ribonuclease digestion. In one mRNA there may be more than one polyA addition signal that are differentially used by tissues. Example is to specify the form of immunoglobulins produced being soluble or membrane-bound. Differential usage of the polyadenylation sites could also determine the intracellular location of the mRNA and hence the protein, e.g., location of a protein in cell body vs. neurites of a neuron. Other functions of the polyA tail are; it protects 3'-end of mRNA from 3'⇒5' exonuclease through binding og the poly(A) tail to cytoplasmic poly(A) tail binding proteins; facilitates mRNA transport into cytoplasm, and control translation initiation. With the exception of histone mRNAs, all mRNAs have the poly (A) tail. Mutations in the polyadenylation sites prevent proper mRNA processing and shorten its half-life. They are implicated in a number of diseases, e.g., some forms of β-thalassemia and hereditary thrombophilia. The synthesis of the poly(A) tail is inhibitable by the terminating 3'-deoxyadenosine (cordycepin) because polymerization cannot proceed without the 3'-OH group.

*Worth Noting: 5'- and 3'-Untranslated regions of mRNA*

After all of these processing, still the mRNA contain long 5'- and 3'-untranslated regions (UTR) that have protein-binding regulatory sequences to modulate mRNA processing, transport, translation and half-life. These UTRs form specific secondary structures, e.g., stem-loop structures that affect the usage and half-life of the mRNA.

- *Intron removal (cutting followed by splicing;* collectively called *Splicing):* It is the removal of large intervening internal protein non-coding RNA sequences from the primary RNA transcript, i.e., introns; and, splicing (joining) the translatable sequences, i.e., exons (EXpression ON). The process is catalyzed by the RNA-protein-enzyme complex called spliceosome. Most splicing occurs after completion of transcription, capping and polyadenylation tailing particularly for short transcripts. However, spliceosome bound to the C-terminal domain of the RNA Pol II during transcription removes some introns cotranscriptionally particularly for large complex transcripts. A gene may contain 1 - 80 introns. Essentially all human genes contain introns except the histone and the rRNA genes. Introns separate functional domains of the gene (exons) that in turn encode distinct functional protein domains. Introns also allow recombination between exons of different genes and the alternative splicing of exons of one gene to produce more than one protein from one mRNA, i.e., exon shuffling. Introns may also contain gene expression regulatory sequences. Except separating functional domains, most introns have no other known cellular functions, but a few may encode functional RNAs or proteins. Failure of the removal of introns - due to mutation of the intron-exon boundaries - prevents mRNA maturation and transport to the cytoplasm; and the mRNA is degraded in the nucleus and no protein will be produced from its translation; i.e., it leads to a gene loss-of-function.

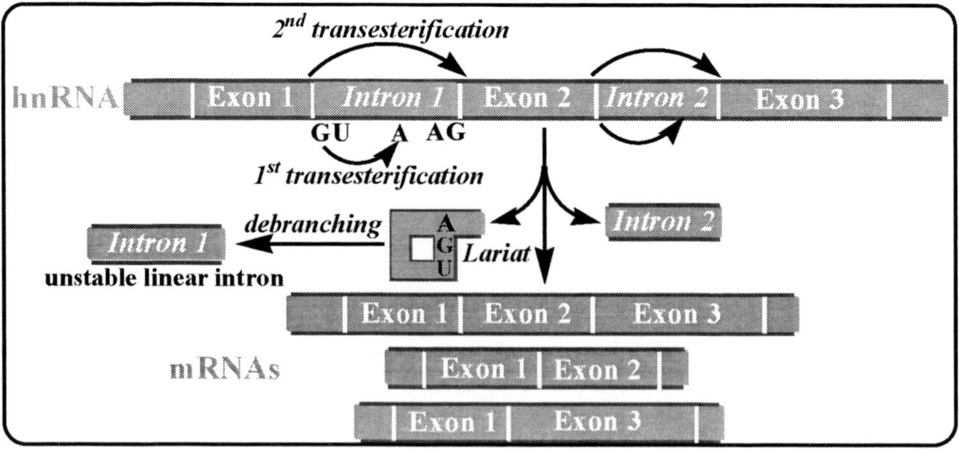


Figure 53. Intron removal maturates mRNA and allows exon joining and shuffling.

*Spliceosome* is an aggregation formed in the nucleus of the substrate *heteronuclear RNA (hnRNA;* as the blue script of mRNA or tRNA) + *4 small nuclear RNA (snRNAs; U1, U2, U5 and U4/U6)* + more than *60 proteins*. Intron removal is formed of a nucleolytic reaction followed by a ligation splicing reaction – in a two trans-esterification reactions: 5'-G of intron into 2'-adenosine branch point and 3'-end of exon 1 into 5'-end of exon 2 with the help of complementary sequences in the snRNAs; see Figure 53. The produced linear introns are degraded by exosome (an aggregate of 3'- and 5'-exonuclaeses and RNA helicase) while the cap and poly(A) protected mature mRNA are not degraded.

There are highly splice-conserved consensus sequences at the 5'- and 3'-ends of essentially all intron-exon boundaries, e.g., 5'-GU-AG-3' and a branch point adenosine 20 - 50 bases from the downstream 3'-splice site. The enzymatic activity of the spliceosome complex resides in the snRNAs that are hence called ribozymes. The snRNAs bind each end of the intron by base pairing with each other, remove the intron and rejoin the ends of exons. Many introns are self-splicing, i.e., they accurately and efficiently splice themselves without requiring additional protein factors. Moreover, 5'- and 3'-splice conserved consensus sequences could be located at different places within an exon leading to alternative retention of the first or the last portion of that exon. Among the other mechanisms of RNA alternative splicing is through alternative retention of an intron. The gene may have a cassette exon that could be omitted. In another case, the gene may have mutually exclusive exons, i.e., incorporation of one in the mature mRNA exclude incorporation of the other.

Because of splicing, the final size of mature mRNA does not correlate directly with gene size or number of exons or the final molecular weight of the protein. Therefore, only 5% of nucleotides transcribed by human RNA Pol II are retained as mature mRNA. Example is the dystrophin gene on chromosome Xp21.2 with 2400 kb genomic size; 79 exons/78 introns giving rise to its largest mature mRNA of only 14,069 bp in size and the largest protein isoforms of 3684 amino acids (427 kDa MW). Introns constitute 98% of the total gene size. The several isoforms of dystrophin are due to tissue-dependent alternative promoter usage; alternative exon splicing and/or alternative usage of polyA addition signals.

*Worth Noting: Inherited defect and splicing*

Inherited defect in splicing due to exon-intron junction mutation may lead to a disease, e.g., at least one form of β-thalassemia due to absent β-chain of hemoglobin. Patients suffering from systemic lupus erythematosis have auto-antibodies against several targets including U1 snRNA of the spliceosome.

- *Editing of mRNA:* Minimal chemical modifications, e.g., methylation occur in pre-mRNA. These post-maturation chemical modifications that have the potential to change the coding properties of mature mRNA sequence after its complete processing and transfer into cytoplasm re called mRNA editing, , i.e., mRNA editing,. They cause subsequent change or truncation in the protein molecule. Example is the apolipoprotein *Apo-B100* (a single polypeptide formed of 4536-amino acid residues) is the product of the complete mRNA expressed in liver, whereas, the *Apo-B48* (a single polypeptide formed of 2152-amino acid residues that is 48% of the apoB100 size) is the truncated form of the same mRNA expressed in intestinal mucosa. The intestinal form of mRNA is produced by deamination of cytosine at **C**AA codon of the glutamine residue number 2153 into uracil by intestine-specific cytosine deaminase. The newly formed UAA is a new upstream stop codon that leads to truncation of the B100 into B48 size upon translation. Editing also helps diversitize the human immunoglobulins.

## Post-transcriptional Processing of tRNA

Processing of the pre-tRNA includes three major modifications:

- *Cleavage:* Primary tRNA transcript is a large precursor containing more than one tRNA unit. It is cut by the action of specific class of ribonucleases. Single unit folds onto itself into clover-leaf shape with a unique secondary structure, various domains and secondary loop structures. Introns - sometimes present - are excised subsequently by specific splicing endonucleases, the generated two halves are joined, and, 5'- and 3'-extra sequences are removed - in a process requiring GTP and ATP hydrolysis. The highly conserved 5'-CCA-3' terminus is added in place of the 3'-UU; see below. All processing occur in the nucleoplasm. In rapidly growing cells tRNA constitutes 15% of the total RNA.
- *Nucleotide chemical modification:* Over 50 different nucleotide chemical modifications are known to occur to tRNAs and rRNAs. These chemical modifications modify the structural and catalytic feature of individual tRNAs/rRNA and their interaction with enzymes and proteins. They also help one tRNA to recognize the several degenerate codons of one amino acid (Wobbling). Chemical modifications include: methylation, deamination of adenosine into inosine, $O_2 \Rightarrow S$ replacement to produce 4-thiouridine, base isomerization by changing position of the ring atoms converts uridine into pseudouridine, and replacement of regular nucleotides with rare one, e.g., queosine and wyosine.
- *Acceptor terminus:* The 3'-terminals UU of the tRNAs are replaced by the characteristic amino acceptor 5'-CCA-3' terminus by the enzyme nucleotidyl transferase.

## Post-transcriptional Processing of rRNA

There are hundreds of copies of rRNA genes arranged tail-to-head in tandem on the nucleolar satellites of the acrocentric 4 chromosomes 13, 14, 15, 21 and 22 - to provide the enormous number ($10^7$) of ribosomes required for each cell. In rapidly growing cells rRNA constitutes 80% of the total RNA. The primary large 45S precursor intron-less transcript (13.7 kb) is cleaved by specific endonucleases and exonucleases into 5.8S, 18S, and 28S rRNAs; see Figure 54. The 5S rRNA is transcribed from a separate nuclear gene. Individual units are subjected to several types of chemical modifications at their nucleotides (see tRNA above). Ribosomal proteins are import into the nucleus from cytoplasm (~560,000 ribosomal proteins per minute are imported into the nucleus with active transport) to assemble with rRNAs in the nucleolus (including the 5S rRNA that comes from nucleus transcribed from a different gene by Pol III) into the small and large ribosomal subunits that are actively exported back to the cytoplasm. The small nucleolar RNA play a major role in such processing.

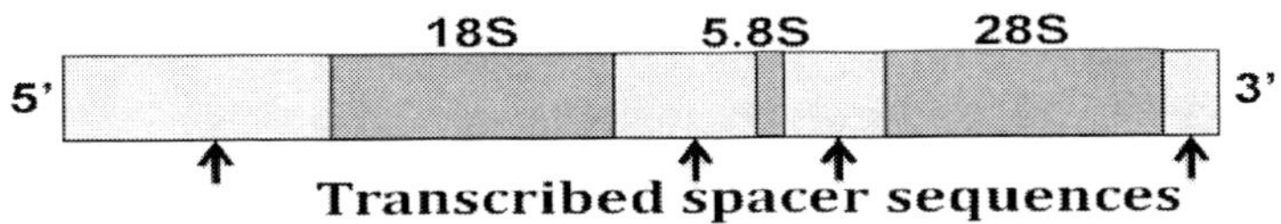


Figure 52. The primary rRNA transcript. The spacer sequences that will be removed to separate the mature rRNAs are shown.

## *Major Differences between Prokaryotic and Eukaryotic mRNA (Table 6)*

In prokaryotes, DNA is not complexes with histones, there are no separation between the circular chromosomal or plasmid DNA and cytoplasm, and, there are no introns in their genes. Transcription and translation are coupled in prokaryotes, where, ribosomes bind and translated mRNA while its transcription is still in progress with minimum - if any - posttranscriptional modification. Moreover, in prokaryotes rate of translation of mRNA affects transcription rate. Prokaryotic mRNAs are very short-living and necessitate continuous transcription to be able to acclimatize to very changing and adverse environment. Likewise, prokaryotic rRNA and tRNA are immediately usable post-transcription. Additionally, bacterial mRNAs are mostly polycistronic, i.e., contain more than one translation unit (a cistron), where, a single primary mRNA transcript is translated into multiple polypeptides because of the presence of multiple independent translation start (AUG) and stop (UAA) codons. Such polycistronic mRNAs are also transcribed from genomes of mitochondrial, slime mold cells and several viruses. However, transcription and translation are uncoupled in eukaryotes due to compartmentation of DNA and transcription in the nucleus and the protein synthesis machinery in the cytoplasm. RNAs including mRNA require extensive posttranscriptional modification. Most of the RNAs have relatively longer half-life and are monocistronic.

**Table 6. Major differences between prokaryotic and eukaryotic mRNAs**

| | Prokaryote | Eukaryotes |
|---|---|---|
| Half-life. | Short | Wide range |
| Number of cistron. | Polycistronic | Monocistronic |
| Presence of introns and their splicing. | No | Yes |
| Separation of translation from transcription. | No | Yes |
| Translation during transcription and their regulatory interaction. | Yes | No |
| Posttranscriptional processing of mRNA (capping/tailing/editing). | No | Yes |
| Processing of large precursor rRNA and tRNA into smaller units. | Yes | Yes |

# Regulation of Gene Expression

On an average, a bacterial genome would have about 4,000 genes and the human genome perhaps 25,000 genes. Analysis of the different type of proteins in a specific type of cell at specific metabolic state indicates that only a fraction of these genes are expressed at any given time. In a bacterium, some gene products are present in very large amounts, e.g., the elongation factors required for protein synthesis, whereas, other gene products occur in much smaller amounts; e.g., a few molecules of the DNA repair enzymes per bacterium.

Requirements for some gene products change over time. The need for enzymes in certain metabolic pathways may increase and decrease as food sources change or are depleted. The organism and hence the cells have to respond to various endocrine and environmental stimuli by changing the rate of expression of certain genes. During embryogenesis of a multicellular organism, some proteins that influence cellular differentiation are present for just a brief time in only a few cells. Differentiation of the cells in various tissues changes the expression pattern of their genes. An example is the uniquely high concentration of a single protein, hemoglobin, in reticulocytes and red blood cells.

Some of the genes products are required at all times during the life of the cells, such as those for the enzymes of central metabolic pathways and cytoskeletal proteins. They are expressed at a more or less constant level in almost all the cells of an organism. Such genes are called as housekeeping genes and this constant expression of a gene is called constitutive gene expression. Their regulatory sequences are usually as simple as a promoter.

The expression of other genes that respond to the metabolic, endocrine, cellular homeostasis or environmental factors changes as per the existence and balance of these stimuli. The cellular levels of products of these genes rise and fall in response to molecular signals. These genes are under regulated gene expression. Gene products that increase in concentration under particular molecular circumstances are referred to as inducible; the process of increasing their expression is induction. The expression of many of the genes encoding DNA repair enzymes, for example, is induced by high levels of DNA damage. Conversely, gene products that decrease in concentration in response to a molecular signal are referred to as repressible, and the process is called repression. For example, in bacteria, ample supplies of tryptophan lead to repression of the genes for the enzymes that catalyze tryptophan biosynthesis.

Understanding the control of the different mechanisms of gene expression is fundamental to development, the utmost economical cellular homeostasis, explaining the related genetic diseases including inborn errors of metabolism, diabetes, cancer, autoimmune diseases and aging, and mechanisms of certain drug action. The expression of various genes in a cell has to be tightly regulated at least at five levels:

- At the epigenetic level
- By gene rearrangement
- By utilization of alternative promoters
- At the transcription level. It is the only functional mechanism in prokaryotes that is control gene expression that is mainly through the fed-starvation state
- At the post-transcription level.

## Epigenetic Control of Gene Expression

An epigenetic control is a heritable and modifiable change in the genomic DNA that controls the rate of gene activity and expression but does not involve a change in the DNA sequence. This kind of control is essential for and explains the cellular development changes, cell differentiation, cell-environment interaction, and, pathogenesis, prognosis and therapeutic outcome of several diseases. Thus, epigenetics bridges the gap between genotype and the phenotype.

Epigenetic factors that control gene expression include:

1. *The spatial cell-cell and cell-matrix contact.* Contact inhibition of cell proliferation and change of the nature of cell differentiation on specific extracellular matrix are examples.
2. *Methylation activated by DNA methyltransferase at cytosine of specific 5'-CpG-3' islands* of the promoter region of genes is a very important regulatory mechanism. *Hypomethylation activates, whereas, hypermethylation silence gene expression through interaction with histone deacetylases*. The methylation pattern is not random and aims at maintaining the paternal pattern of methylation in the newly synthesized DNA after replication, or to acquire new methylation patterns. The preferential expression of a particular gene exclusively from the allele contributed by one parent is called *parental imprinting.* At least thirty human genes display imprinting. DNA methylation is under hormonal regulated.

   Housekeeping genes have unmethylated CpG islands, whereas tissue-specific genes are unmethylated only in target tissues, a pattern inherited from mother cell. Early during embryogenesis, one of the two female X chromosomes is totally silenced by acetylation/Xist RNA association/histone replacement/methylation and appears as a condensed as *Barr body* in the nucleus, which is comprised entirely of heterochromatin. In a cell undergoing X inactivation, the specific inactivation center on one of the X chromosomes starts forming heterochromatin that spreads out to the entire chromosome but leaves a few short segments containing small clusters of genes that remain active. The expression from the single active X is upregulated in both males and females to match the rate of expression from the autosomes. DNA methylating alkylating agents could corrupt gene expression through modulation of the normal pattern of methylation. Hypermethylating inactivation of the cell cycle inhibitor p16 protein in several types of cancers and specific type of retinoid receptors (RARβ) in bladder cancer are examples of disease-implicated abnormal hypermethylation.

*Worth Noting: DNA methylation and genetic defects*

*Prader-Willi* and *Angelman syndromes* are examples of epigenetic diseases caused by differential inactivation by methylation and/or deletion in the mother vs. father genes on chromosome 15. Hypo- and hypermethylation cause altered gene expression as seen in *Immunodeficiency-Centromere instability and Facial anomalies syndrome* and cancer, respectively. Environmental agents, e.g., alkylating pollutants may work by modulation of such methylation and DNA remodeling.

3. *Chromatin remodeling:* Nucleosome and higher chromosome organization and subsequent function may vary with covalent modification of the nucleoproteins. Core histone proteins are susceptible to acetylated (by histone acetyltransferases) or deacetylated (by histone deacetylases) at their amino terminal tail lysine and histidine amino acids to decondense chromatin, recruit coactivators and induce transcription, or, recruit corepressors, condense chromatin and suppress transcription, respectively. Histones are also susceptible to methylation/demethylation, phosphorylation/ dephosphorylation and ADP-ribosylation. Facultative heterochromatin is an example of reversible chromatin remodeling.

## Control of Gene Expression through Gene Rearrangement

The number of copies and position of the different genes helps in controlling their expression. The following events are involved in gene rearrangement.

1. *Chromosomal recombination (cross-over)* is a recombination between homologous metaphase chromosomes during meiosis or between sister chromatids during mitosis. With equal reciprocal exchange of genetic material, it has no pathological consequences. Unequal reciprocal exchange of genetic material results in diseases due deletion/insertion mutation, e.g., hemoglobin β-δ Lepore and anti-Lepore. *Chromosomal translocation* is the joining of a large piece of one chromosome to another unrelated chromosome due to abnormal recombination and/or after exposure to ionizing radiation, free radicals and/or chemical carcinogens that create free chromosomal end. This is frequently observed in cancer cells and has very deleterious effects by merging genes, parts of genes, and/or brining abnormal gene expression sequences; all could inactivate an antioncogene or activate an oncogene.
2. *Differential gene rearrangement:* A number of genes exist as multiple copies and recombination of "xth" copy of 'X' gene with "yth" copy of 'Y' gene results in the production of the functional protein. This kind of rearrangement is the basis for the diversity and polymorphism of proteins like immunoglobulins. There are large number variable sequences for each of the V (hundreds), D (20 types) (20 types), J (6 types) and H (9 types) segments of the heavy- and light-chain genes but are widely separated in the chromosome. During the maturation of the plasma cell, recombination among one V = one D + one J creates a massive number of combinations upon their fusion together with one of the 9 heavy chain elements into the same transcription unit to be express into one mature mRNA molecule.

*Worth Noting: Hypermutation*

*Hypermutation* occurs when a cell allows the rate at which mutations occur in its genome to increase. This is a part of the mechanism used to diversify immunoglobulins along with the genome rearrangements. Hypermutation of the V-gene segments after assembly of the intact immunoglobulin gene is due to mutation rate 6 - 7 orders of magnitude greater than the background mutation rate in the rest of the genome. This is due to the unusual behavior of the mismatch repair system using the daughter rather than the parent strand as a reference, and so stabilizes the mutation rather than correcting it.

3. *Transposition:* Transposons are small mobile DNA sequences that have the capability of changing their position in the genome (transposons or jumping genes) and are found in all organisms (viruses, bacteria and eukaryotes). They may affect the function of the neighboring sequences and may transfer flanking sequences of DNA that profoundly affect the development. They may utilize transposase - a similar enzyme to the retroviral integrase for integration. Integration occurs directly or through an mRNA-cDNA-dcDNA intermediates produced by their reverse transcriptase activity since these transposons (~5 kb in length) contain central protein-coding sequence that give rise to homologous proteins to that of retroviruses, e.g., reverse transcriptase and integrase. Examples of those using an mRNA intermediate include the inactive pseudogenes for α-globin and immunoglobulins, which are processed genes (no introns and they have polyA tail and 5'-untranslated sequence) inactivated by nonsense mutations. They also have end sequences similar to transposons. Transposons also include the Alu family of moderately repeated sequences that are capable of transposing themselves in and out of the host genome by having LTR and integration end sequences similar to retroviruses (e.g., HIV) that help integration.
4. *Gene conversion* is due to abnormal recombination of similar DNA sequences on homologous or non-homologous chromosomes to be in one chromosome. The elimination of any mismatched sequences and repeated similar sequences lead to *gene deletion* and allelic loss. Examples include deletion of tumor suppressor genes during carcinogenesis.
5. *Integration of oncogenic viruses:* Oncogenic DNA and RNA (after conversion into dscDNA) viruses integrate into host DNA by homologous recombination or by protein directed site-specific integration. The expression of the neighboring genes may be altered due to their proximity to viral regulatory sequences.
6. *Gene amplification:* Gene amplification, increasing the number of copies of a gene is another regulatory mechanism for adaptation to the physiological and pathological changes. This happens upon repeated initiation during DNA replication that increases the gene copy number and dosage, e.g., genes for eggshell formation are amplified during oogenesis. A number of genes are amplified in the cancer cells, e.g., amplification of dihydrofolate reductase in methotrexate-treated cancer cells.

## Control of Gene Expression through Utilization of Alternative Promoters

This is a type of regulation of tissue-specific gene expression that is achieved by usage of different elements of the promoter or usage of different promoters; and, consequently an alternative reading frame. An example is the glucokinase gene expressed in the liver and β-cells of pancreas. In both, the gene consists of 10 exons. In liver, the utilized promoter is close to exon 1 and during splicing both the promoter and exon 1, sequences are removed. In β-cells, the utilized promoter is far upstream exon 1 and during splicing exon 1 is joined to other exons of the gene and is not removed. Other than causing structural and functional differences of the protein, this differential promoter usage causes differential control of rate of transcription of the gene in the two tissues.

## Control of Gene Expression at the Transcription Control Level

Transcriptional control through induction, repression and/or attenuation of the rate of mRNA transcription is the sole prokaryotic mechanism of gene expression control that is just one among several mechanisms in eukaryotes. Regulation of gene expression at the transcription level is best illustrated by the *Lac-Operon* in *Escherichia coli* (*E. coli*) as the simplest model of a regulated gene expression, since regulation of other genes in *E. coli* or eukaryotes are far more complex. Under normal growth conditions of *E. Coli*, the cell contain only 600- 800 proteins although it is genetically having thousands of gene, where, the rest of gene are turned off in response to the environmental needs.

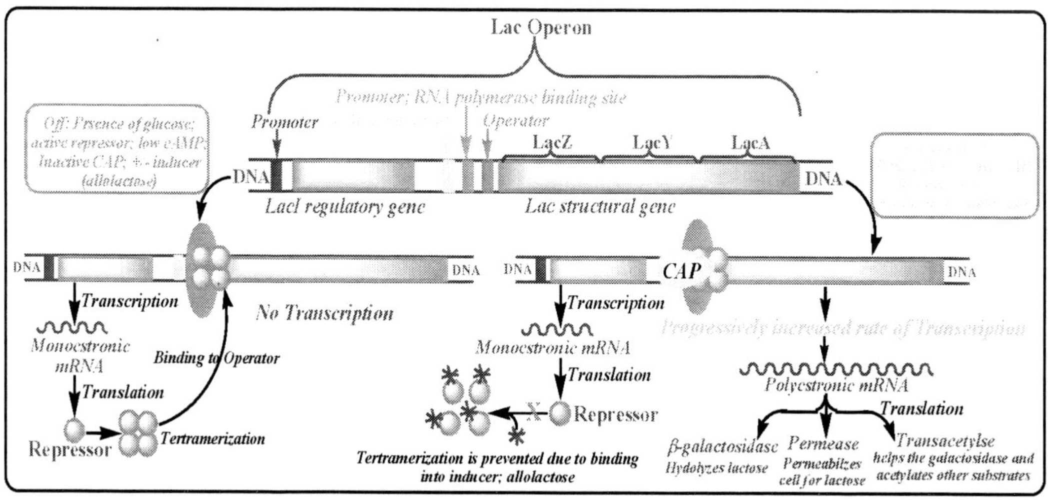


Figure 55. The structure and mechanism of the control of expression of the lac-Operon in E. coli. The Off-state is illustrated on the left side and the On-state is illustrated on the right side. CAP = Catabolite gene Activator Protein.

The genes encoding proteins involved in a related function (e.g., one metabolic pathway) are often grouped sequentially in units called *Operon*. Operon is a coordinated unit of gene expression, i.e., its genes are turned on or off – although they may be separated apart within the genome. Lac-Operon means Lac = lactose-metabolizing enzymes, Oper = operation, and On = on. *Lac-Operon* is formed of two genes; *the regulatory LacI gene* and *the structural inducible Lac-gene*, along with their regulatory sequences. *The regulatory LacI-gene* is constitutively - with a promoter as the only regulatory sequence - transcribed into a monocistronic mRNA that is translated into the *repressor protein*. The repressor protein binds the operon regulatory sequences of the structural inducible Lac-gene and inhibits its transcription. *The structural Lac-gene* is an inducible gene - with a promoter, a CAP-binding and an operator sequences. It is formed of three units (Z, Y and A) and is transcribed into a polycistronic mRNA that is translated into three co-regulated proteins; *β-galactosidase*, *lactose permease* and *transacetylase*. See Figure 55 for the structure of the Lac-Operon and the two modes of its expression. The Catabolite or cAMP gene Activator Protein (CAP)-bind sequence works as an enhancer, whereas, the operon works as a silencer.

*E-coli* can metabolize glucose, glycerol, lactose or galactose as a source of energy with preference of glucose because the glycolytic enzymes are constitutively active. For the integrated metabolism of lactose, the bacterium requires lactose permease that permits entry of lactose into the cells and β-galactosidase that hydrolyzes lactose into glucose and galactose. The transacetylase function is not fully known but it helps the β-galactosidase in hydrolyzing lactose, transacetylates non-metabolizable alternate substrates, e.g., thiogalactoside, and also may acetylate specific proteins and facilitate gene induction.

Jacob and Monod elucidated the mechanism of the coordinated regulation of the expression of the Lac-Operon in 1961. The two states – repression/induction - of Lac-Operon expression are as follows:

1. *Repression (catabolite repression):* When *E. coli* is grown in *presence of glucose* (whether lactose or other sugars are present or not), transcription of the Lac-gene is repressed and hence lactose is not metabolized. The regulatory LacI repressor protein (38 kDa MW) is synthesized at a constant rate. The LacI repressor protein homotetramerizes and binds to its DNA binding site, i.e., the *Operator Sequence*, which is present between the transcription start and the promoter of the structural Lac-gene. Binding of the repressor protein prevents RNA polymerase from transcribing the structural gene because it prevents the binding of the sigma factor. In this state, the repressor is a negative regulator and its DNA binding sequence acts as a silencer sequence. CAP (see below) is inactive because cAMP is very low in the fed state of the bacterium in presence of glucose.
2. *Induction (or derepression): In absence of glucose*, the Lac structural genes are induced after a brief period of the bacterium stopped dividing. Upstream to the Lac-gene promoter sequence there is a specific sequence known as *CAP (Catabolite gene Activator Protein) binding sequence*. This activator protein (22 kDa MW) is also known as cAMP receptor protein (CRP), because it is activated upon binding to cAMP. Glucose starvation of the bacterium leads to an increase in cAMP level (a catabolite). Upon binding to cAMP, the CAP protein homodimerizes to its DNA sequence. Such binding facilitates the binding of RNA polymerase to the promoter sequence and activates transcription by hindering the binding of the repressor to the operator sequence. This induces transcription to an initial low level (~50 fold the basal level). In this state, the CAP is a positive regulator and its sequence works as an enhancer. There are a number of low molecular weight substances that are called inducers of the Lac-gene. They include metabolizable and nonmetabolizable natural and synthetic isomers of lactose, e.g., allolactose (a natural isomer of lactose), lactose or isopropylthiogalacoside (a non-metabolizable lactose analog). They are called inducers because the presence of minimum amount of each of them in the bacterium induces the expression of the Lac-gene. How? They bind the repressor protein and cause inactivating conformational change in it that prevents repressor tetramerization

and lower its DNA binding affinity. This permits a higher rate of expression of the Lac structural genes. The permease produced as a result of this induction further facilitates the entrance of lactose into the cell. Lactose further disfigure the repressor and relief transcription repression into a maximum rate of induction of 1000-fold the basal level within 10 minutes. The non-metabolizable lactose analogs, e.g., isopropylthiogalacoside (IPTG), are called as *a gratuitous inducer*.

Even in the off state of the operon a very low level of the three enzymes exists in the bacterium that allows a very low lactose level to enter. Therefore, dissociation of the repressor from the *lac* operator has little effect on transcription of the Lac structural genes unless CAP-cAMP facilitates RNA polymerase binding and transcription because the wild-type *lac* promoter is a relatively weak promoter. Moreover, increasing intracellular cAMP even in presence of glucose induces the expression of Lac structural genes. CAP interacts directly with RNA polymerase through the polymerase's α-subunit.

*Worth Noting: Regulon and attenuation of transcription*

CAP and cAMP are involved in the coordinated regulation of many operons, primarily those that encode enzymes for the metabolism of secondary sugars such as lactose and arabinose. A network of operons with a common regulator is called a *regulon*. This arrangement, which allows for coordinated shifts in cellular functions, utilizes hundreds of genes. This mechanism is also a major theme in the regulated expression of dispersed networks of genes in eukaryotes. Some operons (tryptophan, histidine, leucine, phenylalanine, and threonine) are rather controlled by *attenuating transcription rate* through binding to the mRNA during transcription. The binding of surplus product induces an inhibitory conformation structure in the mRNA that hinders the progress of the RNA polymerase, and vice versa, low product level induces another facilitative mRNA conformation change that stimulates transcription.

## Postranscriptional Control of Gene Expression

Control of RNA processing (e.g., capping, tailing, editing), control of RNA transport and rate of translation (e.g., proteins utilized in iron absorption, transport and storage) and control of mRNA half-life are other mechanisms for regulation of levels of gene expression.

Alternative splicing of mRNA to generate different mRNAs and hence different proteins from same mRNA is a tissue-specific adaptive and developmental control mechanism that is achieved by:

1. *Alternative usage of termination-cleavage polyadenylation sites*. IgM as an examples has two forms; the $IgM_m$ that is β lymphocyte membrane-bound and the $IgM_s$ that is a secretory form are created by alternative usage of two termination-cleavage polyadenylation sites of one mRNA.
2. *Alternative intron-exon and cleavage-polyadenylation splicing*. Example is calcitonin gene has for 4 exons and two polyadenylation cleavage sites, and is expressed in thyroid C-cell and neurons. In the C-cells, the mRNA is spliced to contain the first 2

common exons, the third calcitonin-specific exon, and, the first polyadenylation site. In brain neurons, the mRNA is spliced to contain the first 2 exons, the fourth calcitonin gene-related protein-specific exon, and, the second polyadenylation site. More than 30% of human genes are subjected to alternative splicing that profoundly affects protein structure/function relationship, and, tissue and developmental distribution. Alternative splicing is subjected to controlling extracellular signalling mechanisms.

3. *Alternative exon shuffling and usage.* α-Topomyosin has 7 different isoforms in 7 different tissues that are created by alternative exon shuffling of one mRNA.
4. *Small and micro interfering RNAs* (siRNAs/miRNAs) specifically double strand to specific mRNA and cause posttranscriptional gene silencing by steric hindrance with translation, de-capping the mRNA and/or activation of mRNA hydrolysis by RNaseH.
5. *The half-life of mRNA:* Physiological and pathological factors affect rate of gene expression by controlling the half-life of mRNA. They may stabilize the mRNA and prolong its half-life or destabilize it and shorten its half-life. Half-life of mRNAs ranges from 10 minutes to years. Binding of regulatory proteins at specific sequences in 5'- and 3'-untranslated regions of mRNA determine its half-life. The half-life of the mRNAs of iron transporters and binding proteins is regulated by the iron-regulated proteins. While, *the cap, the polyA tail and stem-loop secondary structure at 3'- and 5'-UTR* are structural stabilization factors, the *AU-rich sequences at the 3`-UTR*, e.g., AUUUA, are structural destabilizing factors for mRNAs of cytokines and oncogenes. Although histone mRNAs do not have polyA tails, their mRNAs have very stabilizing *stem-loop secondary structure at 3'- and 5'-UTR. De-capping* that controls the half-life of the mRNA and *cytoplasmic localization* signals and nuclear transport proteins in some mRNA are other posttranscriptional regulated processes.

## The Genetic Code and Protein Synthesis

The subsequent stage of the central dogma is the translation of information coded in the mRNA into sequence of amino acids of proteins is known as *Translation*, i.e., translating the nucleotide sequence into amino acid sequence. Proteins are composed of the different combination of the twenty two types of amino acids. However, there are only four types of nucleotides (A, U, G, and C) in DNA/RNA code. If the information for each amino acid is encoded in mRNA then a set of nucleotide sequence should represent an amino acid. If this specifying sequence would be formed of a pair of nucleotides there will be only $4^2 = 16$ codes; whereas, a sequence formed of sets of four nucleotides gives rise to 256 codes. Therefore, the triple letter code is the nature`s solution to the problem of uniquely relating the 4-nucleotide sequence (A, T, G, C) to a suite of 20 amino acids.

Because of the triplet-based genetic code, the 5'-ATGGAAGTATTT AAAGCGCCACCTATTGGGATATAAG-3' sequence could be read starting from the first nucleotide; 5'-ATG GAA GTA TTT AAA GCG CCA CCT ATT GGG ATA TAA G-3' that is call a reading frame. But it is also differently dividable into triplet codes starting from the second nucleotide and readable as; A TGG AAG TAT TTA AAG CGC CAC CTA TTG

GGA TAT AAG; as a second reading frame. Moreover, it could be read using the third nucleotide as; ...AT GGA AGT ATT TAA AGC GCC ACC TAT TGG GAT ATA AG; as a third reading frame. Upon translation into a protein you can expect that the three reading frame will give rise to three different protein sequences.

Therefore, a given DNA sequence on a given strand can theoretically be translated in three different ways or perspectives known as reading frames. Because the double stranded DNA can be read 5'-3' from both strands, a total of six possible reading frames exist for translating a DNA sequence into proteins through the mRNA intermediate. However, with very few exceptions, only one of these six frames is used for any given DNA coding region. This reading frame is called *open reading frame* (ORF) that is an interval of DNA sequence giving rise to an mRNA with a suitable size before the appearance of the first translation stop codon, start of translation, end of translation and 3'- and 5'-UTR containing translation and half-life regulatory sequences. Example of the genes that may have several usable reading frames (*alternative reading frames*), include the production of the cell cycle regulating and apoptosis-inducing anticancer protein $p14^{ARF}$, through the usage of an ***a**lternative **r**eading **f**rame* (ARF) from gene of another cell cycle inhibitory and anticancer protein, the $p16^{Ink4a}$. The use of one or the other of the available reading frames is controlled by cell- and sex-specific antagonizing/agonizing transcription factors.

*Worth Noting: Hargobind Khorana*

An encoding sequence of sets of consecutive three nucleotides would be the best for making the codes because all probable triple alternative combinations of the four nucleotides utilized in DNA or RNA structure give 64 triple codons or possible amino acid names. This was confirmed practically by the pioneering experiments carried out by *Dr. Hargobind Khorana* and his group. They produced the first synthetic gene, the 77bp yeast tRNA-Ala gene and discovered the codes of all the amino acids and established what is known as genetic code dictionary, see Table 7.

Codon is a three nucleotides sequence in DNA that is copied into an mRNA (with U/T replacement), which determines the type and site of amino acid in the translated polypeptide. Codons are read in the 5'⇒3' direction in the mRNA and each codon is identified by base pairing complementarity in an anti-parallel manner with the anticodon of a specific tRNA (3'⇒5' direction) that carries a specific amino acid.

Among the 64 codons there are three codons (UAA, UAG, and UGA) that are called non-sense or translation stop codons because they do not specify specific amino acid (i.e., meaning-less) but signal the termination of translation of mRNA. However, in exceptional cases specified by the need of the organism and the structure of the used mRNA, two stop codons; UGA and UAG could code for the two rare protein amino acids; namely, selenocysteine and pyrrolysine, respectively. The remaining 61 codons are divided unevenly on the remaining major protein 20 amino acids.

**Table 7. The mRNA genetic code dictionary and the corresponding amino acid or translation stop codons**

| 1st nucleotide | 2nd nucleotide | | | | 3rd nucleotide |
|---|---|---|---|---|---|
| | U | C | A | G | |
| U | Phenylalanine<br>Phenylalanine<br>Leucine<br>Leucine | Serine<br>Serine<br>Serine<br>Serine | Tyrosine<br>Tyrosine<br>*Stop*<br>*Stop* (Pyrrolysine; Pyl) | Cysteine<br>Cysteine<br>*Stop* (Selenocysteine; Sec)<br>Tryptophan | U<br>C<br>A<br>G |
| C | Leucine<br>Leucine<br>Leucine<br>Leucine | Proline<br>Proline<br>Proline<br>Proline | Histidine<br>Histidine<br>Glutamine<br>Glutamine | Arginine<br>Arginine<br>Arginine<br>Arginine | U<br>C<br>A<br>G |
| A | Isoleucine<br>Isoleucine<br>Isoleucine<br>Methionine | Threonine<br>Threonine<br>Threonine<br>Threonine | Asparagine<br>Asparagine<br>Lysine<br>Lysine | Serine<br>Serine<br>Arginine<br>Arginine | U<br>C<br>A<br>G |
| G | Valine<br>Valine<br>Valine<br>Valine | Alanine<br>Alanine<br>Alanine<br>Alanine | Aspartate<br>Aspartate<br>Glutamate<br>Glutamate | Glycine<br>Glycine<br>Glycine<br>Glycine | U<br>C<br>A<br>G |

## Characters of the Genetic Code

The genetic code dictionary shown in Table 6 includes all known amino acid- and stop-specifying codons in all known organisms. However, in every specifies there is preference for specific codons for specific amino acids, i.e., for example the 6 arginine-specifying codons will not show up in one species but rather a limited number of them will show in specified species. Therefore, this dictionary differ among different species, i.e., is more limited. Looking at Table 6 below and knowing the natural distribution of these codes, the following are the major 6 characteristics ascribed to the genetic code that are versatile to its functions:

- *Specific or unambiguous:* Each specific codon indicates a specific amino acid, e.g., UUU encodes phenylalanine only. No two amino acids can share the same codon.
- *Degenerate:* One amino acid may have more than one codon because there are 61 encoding codons that are not evenly distributed on the major protein 20 amino acids. Therefore, one amino acids like arginine would have a maximum of 6 codons (synonyms or nicknames, CGU, CGC, CGA, CGG, AGA and AGG), whereas, another like methionine (AUG) and tryptophan (UGG) would has only one codon. This is why genetic code is said to be degenerate. The number of codons for a particular amino acid correlates with its frequency of occurrence in proteins with lowest being methionine and tryptophan, and, highest being arginine, leucine and serine. Degeneracy minimizes the deleterious effects of mutations. Moreover, some species have a marked preference for certain codons.
- *Universality:* The genetic code is universal, i.e., specify the same amino acid in all living organisms from viruses, bacteria, plants, insects to mammals. However, exceptions occur in *mitochondrial* genome where AUA specifies methionine instead

of isoleucine and UGA specifies tryptophan instead of being a stop or non-sense codon in *nuclear systems*. Also, the nuclear arginine codons; AGA and AGG are stop codons in mitochondrial genome. Some other exceptions occur in ciliated protozoa and the single cell plant acetabularia.

- Non-overlapping: The mRNA codons are read form a fixed start point in a continuous manner in the 5'⇒3' direction without overlap usage of the three-base sequence, i.e., no base functions as a common member of two consecutive codons. *Therefore, addition or removal of a single base leads to shifting of the reading frame and producing totally different amino acid sequence after the shift.* However, in small viruses, their small genome in used in an overlapping manner in every direction possible to get the maximum number of genes possible. In human, rarely genes are overlapping (e.g., members of HLA complex at 6p21.3), and genes sometimes occur within genes such as the snoRNA which are found in nucleolar ribonucleoprotein genes. However, once the boundaries of a gene is specified, the codons within are read for replication and transcription in a non-overlapping manner.
- *Comma-less:* Once the reading is commenced at a specific codon, the sequence is read in a continuous manner without interruption till we reach the stop codon in mRNA. It means that the genetic code is read without any punctuation markers among codons of mRNA and DNA, i.e., there are no codons or nucleotides representing a 'comma' or 'conjunction' etc.
- *Colinearity of gene and product:* There is a linear correspondence in base sequence in the gene, its mRNA (after splicing) and amino acid sequence of the produced polypeptide.

## Crick's Wobble (Shaking) Hypothesis

A closer look at the genetic code dictionary reveals that multiple codons for the amino acids generally differ in their third nucleotide. The reduced specificity at the third position in the codon is known as *third base degeneracy* or *Wobbling phenomenon* because it does not obey the strict rule of complementarity upon base pairing with the tRNA anticodons. This enables one tRNA with a specific anticodon to base pairs to several codons of one amino acid that differ only for the 3$^{rd}$ position nucleotide since codon-anticodon binding at 3$^{rd}$ base is not very strict.

The anticodon of tRNA identifies a number of synonym codons of one amino acid that differ at the 3$^{rd}$ base. If the 1$^{st}$ base of the anticodon is C or A, it pairs respectively with G and A only at 3$^{rd}$ base of the codon; but if it is U, it can pair with either G or A as 3$^{rd}$ base of the codon; if it is G, it can pair with C or U as 3$^{rd}$ base of the codon; and if it is inosine 'I' (nitrogen base as hypoxanthine), it can pair with A, C or U as 3$^{rd}$ base of the codon; see Table 8.

Examples of wobble codons are the 2 arginine codons; AGA and AGG that bind same tRNA-UCU anticodon and the three glycine codons; GGU, GGC and GGA that bind same tRNA-CCI anticodon. *Wobbling and hence degeneracy limits the types of tRNA required to translate all the amino acids, therefore, only 33 tRNAs are required for all the twenty two amino acids (including one specific for the initiation AUG). Wobbling also minimizes the effects of mutations.*

**Table 8. Wobbling in the binding of the 1st base of tRNA anticodons and the 3rd base of the mRNA codon**

| 1st Base of the tRNA AntiCodon | Bind With (3rd base of mRNA Codon) |
|---|---|
| C | G |
| A | U |
| U | G or A |
| G | C or U |
| I | A, C or U |

# Protein Biosynthesis (Translation of mRNA)

A polypeptide is a linear amino acid polymer in a specific sequence that is dictated by the code sequence of its mRNA. Proteins are synthesized in cytoplasm, endoplasmic reticulum and/or mitochondria. There are various types of proteins with different structure and function due to the difference in amino acid sequence in their constituting polypeptides. Cell utilizes one or more genes to produce each type of polypeptide.

The prokaryotic protein synthesis has been well studied and has served as a model for the eukaryotic system. *The mechanism of eukaryotic protein synthesis as compared to the prokaryotic system will be described here.*

*Requirements for protein synthesis:* 22 amino acids, tRNAs (33 type), mRNA, amino-acyl-tRNA synthetases, ribosomes, several protein factors, and, ATP and GTP for energy. The process of protein synthesis includes five steps:

- Activation of amino acids
- Initiation
- Elongation
- Termination
- Post-translational modification.

## Activation of Amino Acids

It is catalyzed by amino-acyl-tRNA synthetase that is specialized in binding a specific amino acid to its specific tRNAs by a high energy ester bond, as follows,

*Amino acid + ATP ⇒ Amino-acyl-AMP + PPi ⇒ 2Pi*
*Amino-acyl-AMP + tRNA ⇒ Amino-acyl~tRNA + AMP*

The α-COOH of the amino acid binds to the 2'-OH (in class I synthetases) or 3'-OH (in class II synthetases) group of the ribose of the adenylate of the acceptor arm of the tRNA (i.e.,

3'-ACC-5'). The initiation codon in prokaryotes 'AUG' codes for N-formyl-Methionine (f-met). The formylation of methionine takes place after the amino acid has been attached to tRNA. The enzyme f-Met-tRNA synthetase is different from the normal 'intra-polypeptide' methionine (Met-tRNA) synthetase. In eukaryotes, the initiation AUG specifies the initiation methionine that is carried on a specific initiation met-$tRNA_i$ that is different from internal methionine carrying met-tRNA. Like DNA and RNA polymerases, the amino-acyl-tRNA synthetases and the ribosome have proof-reading characteristics.

*Worth Noting: The second genetic code*

*A "Second Genetic Code":* The enzymes aminoacyl-tRNA synthetases are specific not only for a single amino acid but for certain tRNA as well. The specificity of these enzymes and their importance for protein synthesis has been highlighted by the examples like f-Met-aminoacyl-tRNA and the Seleno-Cysteine-aminoacyl-tRNA synthetases. Selecting the specific tRNA in a matrix of dozens of others is as important for the overall fidelity of protein biosynthesis as the precise selection of a specific amino acid. The interaction between aminoacyl-tRNA synthetases and tRNAs has been referred to as the "second genetic code," reflecting its critical role in maintaining the accuracy of protein synthesis. The "coding" rules appear to be more complex than those in the "first" DNA/RNA code. The tRNA amino acid acceptor arm and the anticodon arm with a specific pattern nucleotide sequence along with the three dimensional structure of the tRNA play an important role in this 'Second Genetic Code' recognition.

The $tRNA^{Sec}$ for 'Seleno-Cysteine – HSe-$CH_2$-$CHNH_2$-COOH' is different from that for serine and specific for selenocysteinyl-tRNA synthesized by the specific seleno-cysteine-aminoacyl-tRNA synthetase. The anticodon of selenocysteinyl-tRNA interacts with a stop codon in the mRNA (5'-UGA-3') instead of a serine codon. The selenocysteinyl-tRNA has a unique structure that is not recognized by the termination machinery and is brought into the ribosome by a dedicated specific elongation factor. An element in the 3'-non-translated region of selenoprotein mRNAs determines whether UGA is read as a stop codon or as a selenocysteine codon. An example of eukaryotic selenoproteins is the antioxidant glutathione peroxidase. The remaining rare protein amino acid; Pyrrolysine (with a pyrroline ring linked to the end of the lysine side chain), is dictated also by a specific $tRNA^{Pyl}{}_{CUA}$ complementary to the 5'-UAG-3' stop codon. A specific aminoacyl-tRNA synthetase transfers pyrrolysine onto its tRNA.

$$\text{(pyrroline ring, N, } CH_3\text{)}-\underset{\underset{O}{\parallel}}{C}-NH-CH_2-CH_2-CH_2-CH_2-\underset{\underset{NH_2}{|}}{CH}-COOH \qquad HSe-CH_2-\underset{\underset{NH_2}{|}}{CH}-COOH$$

**Pyrrolysine** **Selenocysteine**

### *i) Initiation*

Initiation of protein synthesis involves the identification and binding of the mRNA by ribosomes and the binding of the fist initiation amino acyl-tRNA. Requirements for the initiation step include:

i. Initiation amino-acyl-tRNA.
ii. Ribosome.
iii. The mRNA
iv. GTP and ATP.
v. At least 12 eukaryotic (eIFs) but only 3 prokaryotic translation initiation factors.

- *Formation of the 43S preinitiation complex:* The inactive 80S ribosome is dissociated into its subunits with the help of eIF-6 that remains associating the 60S large subunit. Subsequently, eukaryotic initiation factors 1A and 3 (eIF-1A and eIF-3) binds to the 40S subunit to maintain it dissociated and facilitate subsequent steps. The eIF-3 prevents premature association with the large 60S subunit.

  The initiating methionine amino-acyl-tRNA (met-tRNA$_i$) interacts with initiation factor-2 (eIF-2), which has a binding site for GTP. Association of (GTP + eIF-2 + met-tRNA$_i$) enables tRNA to bind the 40S-eIF-1-eIF-3 complex at the P site to give rise to the 43S preinitiation complex. Since the translation always begins at the initiation codon 'AUG', the newly synthesized proteins always start with methionine (or NH-formyl-methionine in prokaryotes) as the N-terminal amino acid. This initiating met-tRNA$_i$ is different from internal met-tRNA.
- *Formation of the 48S preinitiation complex:* Initiation factor-4 (eIF-4; the cap binding protein) binds the cap of mRNA and activates it. The eIF-4 reduces 5'-end complex secondary structure of the mRNA utilizing energy from hydrolysis of ATP into ADP + Pi by its ATPase and ATP-dependent helicase activities. The activated mRNA-eIF-4 complex binds the 43S preinitiation complex to form the 48S complex. This ATP energy-dependent process along with the hydrolysis of the GTP on eIF-2 into eIF-2-GDP ends the formation of the 48S complex with the release of eIF-1A, eIF-2-GDP, eIF-3 and eIF4.

  Upon binding, the 43S complex scans mRNA for the first initiation codon 'AUG' downstream the Cap - using the met-tRNAi anticodon as a hand. This initiation AUG codon differs from the internal AUG in having a specific sequence flanking it known as the Kozak sequence (5'-GCC A/G CC *AUG* G-3'). Thus, anticodon of met-tRNA$_i$ comes in contact with translation initiation codon in mRNA. Very rarely, the GUG and CUG codons encoding valine and leucine, respectively code for methionine when they are located within the Kozak sequence in place of the regular AUG. The cap and the poly(A) tail work synergistically during this step, through a poly(A) binding protein. Since there is no cap in prokaryotic mRNA, the specific *Shine-Dalgarno sequence* upstream of initiating AUG binds to complementary sequence in 16S rRNA and helps identifying the initiation AUG.
- *Formation of the 80S initiation complex:* The GTP-bound initiation factor-5 (eIF-5-GTP) activates binding of the 48S preinitiation complex to 60S ribosomal subunit with the release of eIF-6 and itself, utilizing energy from hydrolysis of its GTP into GDP + Pi. The complete ribosome contains two amino-acyl-tRNA-binding sites; the central *"P" site* (peptidyl site, i.e., the preceding amino-acyl-tRNA site carrying the growing peptide chain) and to its right, the *"A" site* (i.e., the succeeding or incoming amino-acyl-tRNA site). The first amino-acyl-tRNA will be automatically located at "P" site and the next amino-acyl-tRNA enter at "A" site; see Figure 56. A third site in

the complex on the left of the P site called exit or "*E*" **site** shows up upon elongation, where the first preceding empty tRNA departs the complex.

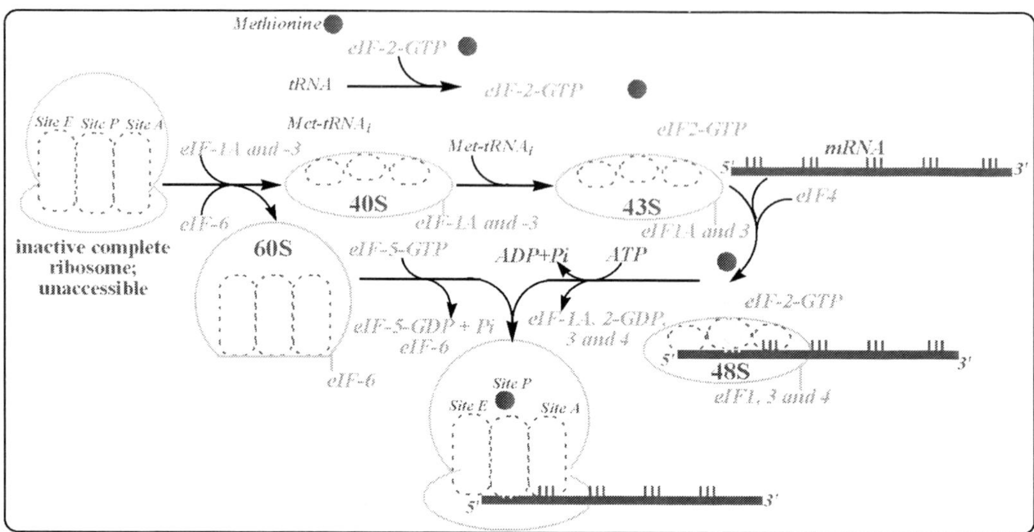


Figure 56. Steps of the translation initiation. The eIF = eukaryotic initiation factor.

### *ii) Elongation*

A translation elongation factor-1α (eEF-1α-GTP) binds and activates the second (or succeeding) amino-acyl-tRNA. This complex allows amino-acyl-tRNA to enter at "A" site of ribosome with the release of inactive eEF-1α-GDP due to hydrolysis of its GTP into Pi catalyzed by an active site on ribosome. The inactive eEF-1α-GDP recycles by activation into eEF-1α-GTP that is catalyzed by its βγ-subunits that help exchanging GDP for GTP; see Figure 57. EF-Tu/Ts functions the same in prokaryotes.

The free α-$NH_2$ group of the 2$^{nd}$ amino acid binds the α-COOH group of the first amino acid (Met$_i$) with the transfer of whole peptide chain to the 2$^{nd}$ tRNA at "A" site by peptidyl transferase (the 28S ribozyme of 60S subunit) with the release of the free initiation tRNA at "P" site.

Translation elongation factor 2 (eEF-2, translocase) moves the whole ribosomal complex, along with the newly formed peptidyl-tRNA, by a one-codon forward along the mRNA in 5' to 3' direction - using the anticodon of the second amino-acyl-tRNA as a second hand. Translocation requires hydrolysis of GTP into GDP + Pi and creates an open "A" site for entrance of a 3$^{rd}$ amino-acyl-tRNA that recognizes the new codon and so no. Polypeptide chain is thus increased by one amino acid each time. Eukaryotic protein synthesis appears to be slower than the prokaryotic one as the eukaryotic ribosome incorporates ~6 amino acids/second, whereas, prokaryotic one incorporates ~18 amino acids/second. One mRNA may be repeatedly translated into up to $10^5$ polypeptide copies before its degradation leading to huge augmentation of the initial signal of a few mRNA copies. The first tRNA - now empty - is translocated at the E site and dissociates. Therefore, while a new hand (anticodon) in budding to catch a new codon on mRNA at the A site, and the second hand is located at the

P site, the third hand - with anticodon of the empty tRNA is retracted at the E site. This enables the ribosome to climb progressively forward along the mRNA.

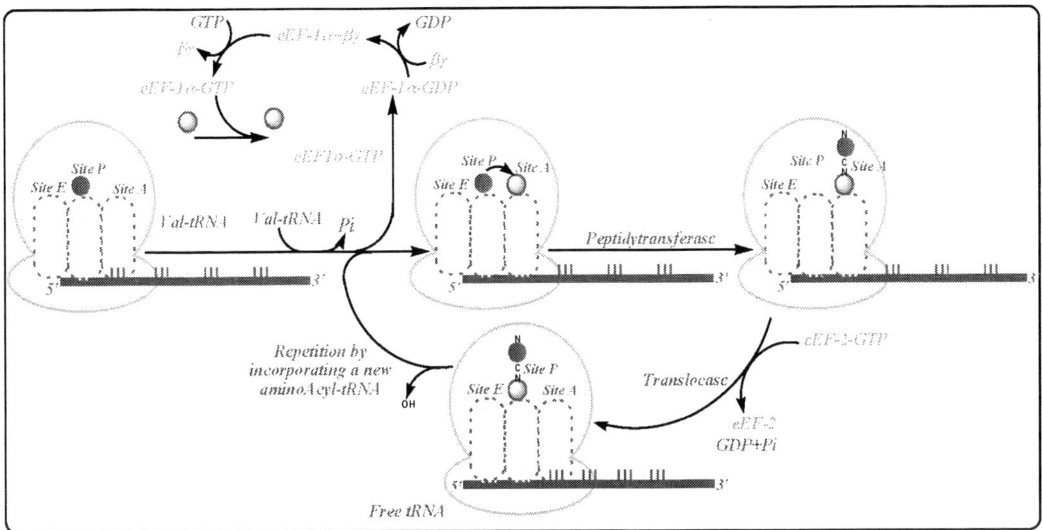


Figure 57. Steps of the translation elongation. eEF = eukaryotic elongation factor.

### *iii) Termination*

When a termination codon in mRNA appears at "A" site, there is no tRNA that can recognize it, but there is the complex of translation releasing factors (eRF1 and 3-GTP) that recognize these codons. A releasing factor activates the peptidyl transferase to hydrolyze and release the peptide chain on last tRNA at "P" site to adopt its three dimensional (3D) conformation and dissociates the complete 80S ribosome into its subunits to release the last free tRNA and mRNA. This requires hydrolysis of its GTP into GDP + Pi.

*Worth Noting: Non-genic proteins*

Rare very essential proteins in human, animal, plant and microbial cells are synthesized denovo, i.e., without an mRNA template or a specific gene for such proteins. The most prominent example is the tripeptide glutathione; see the formula below. Glutathione contains an unusual peptide linkage between the amine group of cysteine and the carboxyl group of the glutamate side chain. Glutathione as an antioxidant protects cells from reactive oxygen species such as free radicals and peroxides. It modulates immune function and participates in several metabolic reaction, DNA repair, protection of enzymes, amino acid transport, and inactivation of insulin. It also, as a nucleophile at its sulfur attacks poisonous and endogenous electrophilic conjugate acceptors. Glutathione is synthesized in two ATP-dependent steps; the rate-limiting γ-glutamyl cysteine synthetase (glutamate cysteine ligase) activates ligation of L-glutamate and cysteine, and, glutathione synthetase that adds glycine to the C-terminal of γ-glutamyl cysteine. All cells are able to synthesize glutathione; however, the most significant tissue is the liver. Mice with genetically lost first enzyme in the liver die within 1 month of birth.

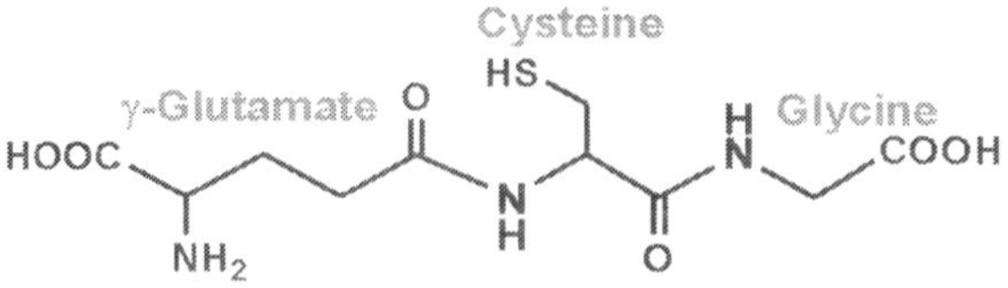

Glutathione - a nongenic tripeptide

## Post-translational Modifications of Proteins

The freshly synthesized proteins need to acquire their native three dimensional conformations by proper folding and aggregation, proteolytic cleavages, and chemical covalent linkages/substitutions to become functional (correct sequence ⇒ correct 3D structure ⇒ normal function). These modifications known as the post-translational modifications are interdependent for the polypeptide to take up its correct three-dimensional conformation without a specific constant sequence of events that applies to all proteins. The posttranslational modifications of the protein start right after it emerges from the ribosome. Furthermore, protein has to be translocated to its proper intracellular/membrane/extracellular target locations, i.e., protein sorting.

- *Proteolytic Cleavage:* The initiation methionine is first cleaved by specific proteases. Proteins with pre-pro-form, e.g., preproinsulin with 105 amino acid residues, undergo further cleavages by removal of the N-terminal signal peptide of 24 amino acid residues. Proteolytic cleavages could occur at both ends of one polypeptide to get shorter or could cut one polypeptide into several functional and/or nonfunctional ones. Conversion of the single polypeptide proinsulin into the dipeptide insulin is due to removal of the internal C-peptide sequence. In some proteins the removal of internal sequences is followed by rejoining of the flanking pieces (this simulates intron-exon splicing of mRNAs).
- *Chemical covalent modifications:* Over 150 types of modifications occur to amino acid side chains or to the amino or carboxyl groups of the terminal amino acids in a polypeptide include: N-terminal lysine residue *acetylation* and *methylation* that is most essential for controlling chromatin function. Proline and lysine in the polypeptide are *hydroxylated* into hydroxy-proline and -lysine essential for the collagen stability. *Carboxylation* of glutamate residues of several blood clotting factors, e.g., prothrombin, activate their function by being able to chelate $Ca^{2+}$. Addition of membrane-anchoring hydrophobic moieties to the proteins include: prenylation (addition of an isoprenoid unit), geranylation (addition of geranyl unit), farnesylation (addition of farnesyl unit), fatty acid acylation, myristoylation, biotinylation and retinoylation (addition of vitamin A or its acid; retinoic acid). ADP-ribosylation on the expense of $NAD^+$ mostly affects negatively the protein function during, e.g., DNA repair.
- However, the most important posttranslational covalent modifications are *phosphorylation* and *glycosylation*. Phosphorylation-dephosphorylation of tyrosine, serine and threonine residues of specific proteins (mostly enzymes) is the most dynamic way of controlling their activity. Glycosylation of membrane and secretory

proteins by addition of carbohydrate groups (onto asparagine N-glycosylation/serine or threonine O-glycosylation, in endoplasmic reticulum and Golgi) gives the protein its identity, stability and sorting signal into lysosome, cell membrane or extracellular space. In case of O-linked sugar chains, they are synthesized by sequential addition of their active nucleotide from. The N-linked sugar branched chains are first synthesis by sequential addition onto the sugar carrier dolichol phosphate from which it is transferred to the N-link. These enzymatic modifications of amino acids and/or their covalent attachment to new chemical groups are essential for: proper folding, the structure-function relationship, protein stability, and function as intracellular and extracellular sorting signals.

- *Proper folding:* The information contained in the primary amino acid sequence of the polypeptide is utilized during the folding process into the correct three-dimensional structure. Proper folding is regulated by *molecular chaperone/chaperonin proteins* that bind and prevent folding and aggregation with other proteins till the whole polypeptide is synthesized. Properly folded protein then assembles with other units and may make super-quaternary aggregation essential to the protein function.

*Worth Noting: Molecular chaperones*

Molecular chaperones are ATP-dependent ATPase proteins that are required for not only proper folding, assembly and aggregation but also translocation of proteins across membranes and degradation of proteins in cytoplasm, mitochondria and ER. These chaperones are induced by stress conditions that lead to mis-folded or damage proteins, e.g., high temperature and various chemicals that alter protein folding. They include heat shock proteins (Hsp) such as Hsp10 and -70 and enzymes such as *protein disulfide isomerase* that forms proper disulfide bonds between inter- and/or intra-chain cysteine residues.

- *Protein degradation:* The concept that the proteome of a cell can change over time requires not only controlled *de novo* protein synthesis but also the controlled rapid removal of proteins whose functions are no longer required. Control of the *protein half-life* is either specified by the nature of its N-terminal amino acid residue or by covalent binding of ubiquitin (i.e., *ubiquitinylation*) to lysine residues as a marker for degradative fate through the proteosome – the protein degradation house. Some amino acids are stabilizing such as serine and methionine, whereas, others are destabilizing and shorten the protein half-life to a few minutes such as arginine and aspartate (Table 9).

*Abnormal cleavages, folding and/or covalent modifications due to afflictions in the processing proteins or due to mutation at the covalent modification site of the polypeptide are implicated in several hereditary diseases and cancer.*

**Table 9. Protein half-life is dependent on the amino terminal amino acid residue**

| Amino acid | Effect | Half-life |
|---|---|---|
| Methionine, glycine, alanine, serine, threonine, or valine. | Stabilizing | >20 hr |
| Isoleucine or glutamine. | Destabilizing | ~30 min |
| Tyrosine or glutamate. | Destabilizing | ~10 min |
| Proline. | Destabilizing | ~7 min |
| Leucine, phenylalanine, aspartate, or lysine. | Destabilizing | ~3 min |
| Arginine. | Destabilizing | ~2 min |

## Pathways of Protein Sorting and Related Diseases (Figure 58)

Proteins synthesized on the ribosomes in cytosol are translocated (sorted) into all the compartments of the cell where they acquire further modifications and function. For sorting, proteins require a built-in structural destination address marker or tag. Protein sorting is a highly complex and specialized phenomenon. Some of the concepts of protein sorting are discussed below along with the disorders that may occur in case the proper sorting of proteins fails.

*Cytosolar protein synthesis and sorting:* The proteins targeted to stay in cytoplasm, mitochondria, nucleus or peroxisome are synthesized on the free ribosomes in the cytoplasm – usually in a polysome complex formed from several ribosomes simultaneously binding and translating one mRNA but in a spaced and placed manner with ~100 nucleotide intervals.

- *Mitochondrial protein sorting:* The mitochondrial proteins are marked by having 20 - 80 amino terminal amino acid leader sequences. These sequences contain basic positively charged amino acids that complex with molecular chaperones and targeting proteins in their unfolded state, a process that requires energy from ATP. Two complex translocases located on outer and inner mitochondrial membranes work with the help of proton-motive force (-ve matrix and +ve inter-membrane space) to translocate these proteins into the mitochondria. *Matrix-processing peptidase* cleaves the leader sequence and then molecular chaperones ensure proper folding and prevent aggregation of the nascent proteins. Proteins may finally reside in the outer-membrane, inner-membrane, the inter-membrane space or the matrix of mitochondria.
- *Nuclear protein sorting:* The proteins meant to enter the nucleus have nuclear localization signal that is rich in basic amino acids. Nuclear targeting proteins like *Importin* and *Ran* interact with the targeting signal. Mostly, the transfer of the proteins across the nuclear membrane is energy consuming and utilizes GTP for the purpose.
- *Peroxisomal protein sorting:* It requires a peroxisomal targeting sequence to import proteins in folded and aggregated state. Zellweger's syndrome is an example of genetic diseases due to defective biogenesis of peroxisomes and its enzymes.

*Endoplasmic reticulum (ER)-Golgi protein synthesis and sorting:* Most of the proteins synthesized on the rough ER-Golgi cisternae system are secretary in nature and hence have to be secreted out of the cells to the extracellular space or blood for body distribution.

They are destined to ER, Golgi apparatus, lysosomes, plasma membrane and secretion in a constitutive transport vesicle or a regulated inducible secretory vesicle. These proteins have a sorting address *signal peptide*; the N-terminal 12 – 35 amino acids sequence that have a hydrophobic central cluster, methionine at the amino end, and a positively charged amino acid close to it. When this sequence emerges from the large ribosomal subunit, it is recognized by *signal recognition particle* (SRP) that is composed of 6 proteins and the 7S RNA. SRP bind also the ribosome and arrests translation elongation at 70 amino acids length. The particle binds its transmembrane receptor on the ER. This receptor works in conjunction with *the translocon* (a protein conducting channel and a receptor for ribosome) to translocate the signal peptide unfolded into the lumen of ER. Translation resumes after that and the signal peptide is cleaved by signal peptidase in the ER. Proteins destined to the ER membrane such as Cyp450 remain attached to the membrane with uncleaved signal peptide as specified by a stop transfer sequence that follows the signal sequence. Other proteins go into the lumen and are N-/O-glycosylated while traversing the ER membrane.

Some proteins and polysaccharides are transported from ER lumen by a retrograde transport system back to cytosol for degradation. Oppositely, some proteins are synthesized on free cytosolar ribosomes and bind spontaneously to ER membrane such as cytochrome b5.

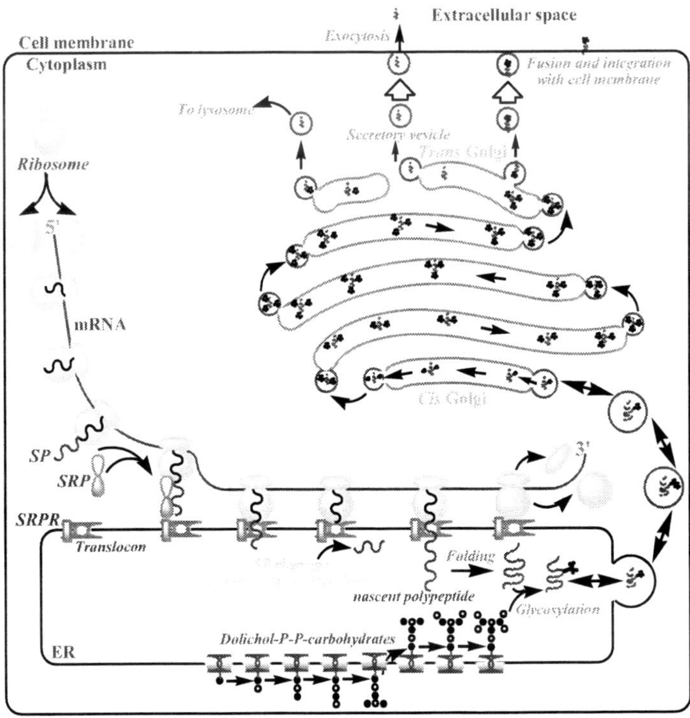


Figure 58. Translation and posttranslational modifications and sorting of the proteins in the endoplasmic reticulum(ER)-Golgi cisternae system. SP = signal peptide; SRP = signal recognition particle; SRPR = signal recognition particle receptor.

Proteins synthesized in the rough ER system are packaged in transport vesicle that buds out from ER and traverse cytoplasm to fuse with the accepting face (the cis) of cisternae of Golgi apparatus for further processing, glycosylation and final sorting into new transport vesicles that bud out of the emitting face of these Golgi cisternae (the trans). Proteins in transport vesicles could be targeted from Golgi back to ER (directed by the signal carboxy terminal lys-asp-glu-leu sequence), to cell membrane structures (directed by the hydrophobic region tags), to lysosomes (targeted through a clathrin-coated vesicle by their mannose-6-phosphate tag that bind specific receptor on lysosome), or fuse with cell membrane for extracellular secretion of its content by exocytosis (no specific tags).

*Worth Noting: Protein sorting and diseases*

Defective inter-compartmental processing and/or sorting of proteins leads to several serious mostly inherited diseases. For example, diseases due to defective lysosomal targeting are called lysosomal storage diseases because wastes accumulate in lysosomes without been cleared by hydrolysis due to mis-targeting and/or deficiency of respective enzyme. Lysosomal storage diseases cause death at early age. *I-cell disease* (*mucolipidosis II*) is an example in which the mannose-6-phosphate tag is missing and lysosomal proteins are sorted instead to the extracellular space. It is due to mutation and deficiency of one of the tag synthesizing enzyme; namely, N-acetylglucosamine phosphotransferase. Lysosomes become stuffed with undigested materials as a lysosomal storage disease with loss of function and severe consequences including death before the age of 8. Other lysosmal storage diseases include gangliosidoses.

## Regulation of Protein Synthesis

There are 5 approaches by which protein synthesis is regulated:

a. *Cellular stresses:* During stresses - such as amino acids and glucose starvation, viral infection, protein misfolding, serum deprivation, hyperosmolality and heat shock, the protein synthesis is halted. The inhibition is brought about by the *inactivation of eIF-2* through phosphorylations by at least 4 different protein kinases.
b. *Inducers:* Since a high level of protein synthesis is required during cell division, the mitogenic factors act also as inducers of protein synthesis. A number of endocrine stimuli also enhance the protein synthesis. Insulin and mitogenic growth factors induce phosphorylation of the inhibitory *eIF-4-binding protein* that releases the free active *eIF-4* leading to enhancement of initiation. This results in the posttranscriptional increase in protein synthesis in liver, adipocytes and muscles. Accumulation of heme - the Fe-binding pigment of the hemoglobin - in reticulocytes (the precursor cell of RBCs) enhances globin synthesis through inhibiting of the phosphorylation of eIF-2 and hence hastens translation initiation.
c. *Regulatory mRNA-binding proteins:* Specific RNA-binding proteins bind to sequences in either the 5'-untranslated region (5'-UTR) or 3'-UTR of certain mRNAs and regulate the rate of their translation. Examples are mRNAs for proteins concerned with iron metabolism. When intracellular iron levels are low the rate of

synthesis of the membrane transferrin receptor mRNA increases so that cells bind plasma transferrin and imports more iron. In the 3'-UTR of the receptor mRNA there is specific iron response element that make hair-pin loop structures recognized by a specific cytoplasmic iron response element-binding protein. This binding happens when cellular iron concentration is low and act to stabilizes mRNA to be available for more translation. When cellular iron level is high, this binding is prevented by iron binding into iron response element-binding protein. The free transferrin receptor mRNA is unstable and gets degraded rapidly and no receptor protein is synthesized. Oppositely, the translation of the ferritin mRNA into the cellular iron sequestrating protein, ferritin, is prevented through interaction of a specific iron response element-binding protein with a specific iron response element at its 5'-UTR. High cellular iron concentration binds into and releases the iron response element-binding protein allowing a high rate of mRNA translation. Low cellular iron allow the iron response element-binding protein to bind and prevents translation of the ferritin mRNA.

d. *Viral and bacterial proteins:* Some viral proteins prevent association of the initiation factors eIF-2 and eIF-4 with mRNA. Diphtheria is a fatal early age disease caused by lysogenic phage infected *Corynebacterium diphtheria* exotoxin. This toxin inactivates eEF-2 by catalyzing its ADP-ribosylation on the expense of NAD and consequently inhibits protein synthesis. This disease is prevented by vaccination and is treated by diphtheria antitoxin that contain antibodies that immunoneutralize the toxin.

*Worth Noting: Effect of antibiotics on protein synthesis*

Many antibiotics that kill pathogenic bacteria work by selectively inhibiting the protein synthesis in bacteria because of the distinction between eukaryotic and prokaryotic ribosomal system.

- Aminoglycosides (e.g., streptomycin), bind proteins and the 16S rRNA of the 30S ribosomal subunit and prevent initiation and causes mRNA misreading into mutant or truncated protein. It is ototoxic and causes deafness.
- Macrolides (e.g., erythromycin), bind the 50S subunit and prevent translocation.
- Tetracyclines reversibly prevent the binding of amino-acyl-tRNA to "A" site in the 30S subunit.
- Chloramphenicol binds the 50S subunit and inhibits peptidyl transferase activity. Mitochondrion is particularly sensitive to chloramphenicol that inhibits its protein synthesis particularly in the bone marrow cells that can cause severe aplastic anemia.
- Because the above antibiotics have the same effect on the eukaryotic mitochondrial system, albeit, at higher concentration, they will have some side-effects on the human body, e.g., fatigue, epigastric distress, diarrhea, and, infrequently, cholestatic jaundice.
- Puromycin has a structure similarity with tyrosyl-tRNA and causes premature release of the peptide in both eukaryotes and prokaryotes. Cycloheximide inhibits peptidyl transferase activity in eukaryotes only. Ricin, a glycosidase that is an extremely potent protein toxin found in castor oil beans. It inactivates ribosomes by cleavage the eukaryotic 28S rRNA and reduces protein synthesis and a single molecule of ricin is enough to kill a cell.

## Review Questions

A. *Write short on each of the following:*
   1. The gene concept.
   2. The gene promoter.
   3. The gene response elements.
   4. The role of mRNA 5'-capping.
   5. Molecular chaperones.
   6. Diseases related to protein sorting.
   7. Post-translational modifications.
   8. Mitochondrial targeting of proteins.
   9. Control of protein half-life.
   10. Transposons.
   11. Gene amplification.
   12. Role of the regulatory mRNA binding proteins in protein synthesis.

B. *Explain each of the following:*
   1. Differences between pro- and eukaryotic RNA polymerases.
   2. Differences between pro- and eukaryotic mRNAs.
   3. The mRNA editing.
   4. The epigenetic regulation of gene expression.
   5. The role of the rho (*p*) factor in transcription.
   6. The Crick Wobble hypothesis.
   7. Genetic code is regenerate.
   8. Regulon and transcriptional attenuation.
   9. Hypermutation.
   10. Lac-Operon in the on-state.
   11. Rifamycin inhibition of transcription.
   12. Why chloramphenicol has side-effects on human.
   13. The effect of cellular inducers and stressors on protein synthesis.

C. *True and false questions:*
   1. Prokaryotic mRNA is monocistronic, whereas, the eukaryotic mRNA is polycistronic. There is no separation between transcription and translation in prokaryotes.
   2. One gene could have more than one promoter and more than one terminator sequences.
   3. The retroviral reverse transcriptase has a high affinity to the dideoxynucleotides than the cellular DNA polymerases.
   4. The specificity of the genetic code means no two amino acids can share the same codon.
   5. Degeneracy of the genetic code minimizes the deleterious effects of mutations.
   6. There are no exceptions for the universality of the genetic code.
   7. The genetic code is overlapping, whereas, genes are non-overlapping.

D. *Supply the Missing Information questions:*
   1. A gene is the smallest physical and functional DNA unit of inheritance that codes for a --------------------------------- .

2. The gene is an integrated functional unit composed of -------------------------, and three types of the transcription regulatory sequences; ----------------------, ----------------- and ---------------------.
3. The gene promoter controls the basal ------------------------------.
4. Initiation of transcription requires the ------------------ factor, whereas, its termination requires the -------------------- factor.
5. Spliceosome is an aggregation formed in the nucleus of the substrate heteronuclear RNA + 4 ---------------------------- + more than ----------------------.
6. Editing of mRNA is ------------------------------------------------ that have the potential to change the coding properties of mature mRNA sequence.
7. Characteristics of the gene promoter include:
   a. It is in a close continuity with ------------------------.
   b. It works -----------------------dependently.
   c. It controls the --------------- level of gene expression.
8. Characteristics of the gene response element include:
   a. It works -------------------------independently.
   b. It is not in a ------------- continuity with the gene proper.
   c. It controls the higher than basal level of ---------------------------.
9. Epigenetic factors that control gene expression include:
   a. The spatial ------------- and --------------------- contact.
   b. DNA -----------------------.
   c. Chromatin --------------------------.
10. Control of gene expression through gene rearrangement include:
   a. Chromosomal ---------------------- (cross-over).
   b. Differential gene --------------------------.
   c. Transposition.
   d. Gene --------------------------.
   e. Integration of -------------------- viruses.
   f. Gene -------------------------.
11. Post-transcriptional control of gene expression include:
   a. Alternative usage of ---------------------------------------------------- sites.
   b. Alternative ------------------- splicing.
   c. Alternative ------------- shuffling and usage.
   d. Small and micro ---------------------- RNAs.
   e. The control of the half-life of ------------- and rate of its --------------------.
12. Cellular stressors inhibit protein synthesis through phosphorylation of ---------.
13. Inducers of protein synthesis work by phosphorylating the inhibitory ------------------------------.
14. Aminoglycosides inhibit prokaryotic protein synthesis by preventing ---------------- and causes mRNA ----------------------.
15. Chloramphenicol inhibits prokaryotic protein synthesis by inhibiting ---------------------------------- activity.

E. *The multiple choice questions:*

1. The lysosomal storage I-cell disease, lysosomal enzymes are found in the blood due to:
   A. Lack of their mannose-6-phosphate.
   B. Non-function proton pump and transport system.
   C. A mutation in the clathrin gene.
   D. Inability to form lysosomal vesicles.
   E. Absence of oligo carbohydrate-side chain.
2. All are required for the binding of RNA polymerase to the prokaryotic promoter sequence EXCEPT:
   A. Rho factor.
   B. the -10 consensus sequence.
   C. the -35 consensus sequence.
   D. Sigma factor.
   E. Helicase.
3. The DNA template strand 5'-GCTATGCATCGTGATCGAATTGCGT-3' gives upon transcription:
   A. 5'-ACGCAATTCGATCACGATGCATAGC-3'.
   B. 5'-UGCGUUAAGCUAGUGCUACGUAUCG-3'.
   C. 5'-ACGCAAUUCGAUCACGAUGCAUAGC-3'.
   D. 5'-CGAUACGUAGCACUAGCUUAACGCA-3'.
   E. 5'-GCTATGCATCGTGATCGAATTGCGT-3'.
4. For lac-Operon, shifting E. Coli into a low glucose medium induces the following sequence of events:
   A. The cAMP level increases, cAMP binds CAP, CAP binds a specific DNA sequence to induce transcription.
   B. The cAMP level increases, cAMP binds CAP, CAP binds a specific DNA sequence to repress transcription.
   C. The cAMP level increases, cAMP binds CAP, CAP binds a specific DNA sequence to repress transcription cAMP level increases, cAMP binds CAP, CAP is released from the DNA sequence to induce transcription.
   D. The cAMP level decreases, cAMP is removed from CAP, CAP binds a specific DNA sequence to induce transcription.
   E. The cAMP level decreases, cAMP is removed from CAP, CAP binds a specific DNA sequence to repress transcription.
5. Heat shock proteins were originally described as proteins produced in response to heat shock. Some are now known to act as:
   A. GTPase-activating proteins.
   B. Protein tyrosine-kinase.
   C. Proteases that degrades ubiquitin-tagged proteins.
   D. Molecular chaperones that regulate protein folding.
   E. Ionophores that dissipates the $H^+$ gradient.
6. Which of the following IS NOT involved in the processing of the eukaryotic pre-mRNA?
   A. Capping of the 5'-end.
   B. PolyA tailing of the 3'-end.

C. Excision of introns.
D. Splicing of exons.
E. Transport of the pre-mRNA into cytoplasm.

7. Which of the following IS NOT a potential problem that could associate cloning and expression of eukaryotic protein-coding nuclear gene in a prokaryotic cell?
   A. Lack of intron-splicing mechanism in prokaryotes.
   B. Difference in translation initiation codons in eukaryotes and prokaryotes.
   C. Difference in post-translational processing in eukaryotes and prokaryotes.
   D. Susceptibility of the protein product to prokaryotic proteases.
   E. Stability of the eukaryotic mRNA in the prokaryote.
   F. Differences in the epigenetic mechanisms in eukaryotes and prokaryotes.
8. During folding of some secretory proteins into tertiary structure, formation of disulfide bonds attain the correct form among several alternatives because:
   A. The unwanted alternative foldings are degraded in lysosome.
   B. Folding and processing are polished in the endosome.
   C. Disulfide bonds are isomerized into correct from in the endoplasmic reticulum.
   D. Only correctly folded proteins are translated in the endoplasmic reticulum.
   E. Arginine guides the correct formation of disulfide bonds.
9. During activation of amino acids into amino acyl-tRNA PPi rather than Pi is produced because:
   A. The hydrolysis of the α-β-phosphodiester bond releases more energy.
   B. Faster binding of the enzyme to βγ-phosphate groups favor the reaction.
   C. AMP produced is readily converted into ATP by adenylate kinase.
   D. Hydrolysis of PPi into 2 Pi favors completion of the reaction.
   E. PPi has a wider buffering action than Pi.
10. In eukaryotes, one of the following DNA gene regulatory sequences works at any location and orientation to induce a high rate of transcription:
   A. Promoter.
   B. Splice donor site.
   C. Enhancer.
   D. Operator.
   E. Corepressor.
11. If all of the cysteine-tRNA is converted chemically into alanine-tRNA and is added to the translation system of an mRNA that contains both cysteine and alanine codons:
   A. Cysteine replaces alanine each time the alanine codon is translated.
   B. Alanine replaces cysteine each time the cysteine codon is translated.
   C. The protein will be alanine deficient.
   D. The protein will have no changes in its amino acid sequence.
12. With high lactose/low glucose conditions maximal transcription activation of the lac operon in *E. Coli* could be induced by:
   A. Induction of the *LacI gene*.
   B. Loss of the CRP activity.
   C. Accumulation of cAMP.
   D. Inducer deficiency.

E. Repressor binding at the operator.

13. Termination of RNA synthesis may be signaled by a sequence that is recognized by:
    A. Rho (*p*) factor.
    B. Sigma factor.
    C. Alpha factor.
    D. None of the above.
    E. All of the above.
14. The binding of prokaryotic DNA-dependent RNA polymerase initiation of RNA synthesis:
    A. Erythromycin.
    B. Streptomycin.
    C. Oligomycin.
    D. Rifamycin.
    E. Heparin.
15. In the classic model of Lac-Operon transcription control a repressor protein bind to:
    A. an enhancer.
    B. an AUG sequence.
    C. an operator.
    D. a TATA box.
    E. a ribosome binding site.
16. One of the following statements best describe the role of the sigma during transcription:
    A. It is essential for elongation of the RNA transcript.
    B. It recognizes and helps RNA polymerase binding to the promoter sequence.
    C. It increases RNA polymerase binding to any DNA sequence.
    D. It is required for termination of transcription.
    E. It prevents the dissociation of the core enzyme during elongation.
17. Which of the following IS NOT a property of the mammalian signal recognition particle?
    A. It targets the nascent secretory polypeptides to endoplasmic reticulum.
    B. It temporarily arrests translation.
    C. It has one molecule of RNA and several proteins.
    D. It has a signal peptidase activity.
    E. It binds the signal sequence of the secretory polypeptides.
18. The ribosome does all of the following EXCEPT:
    A. Peptide bond formation.
    B. Amino acyl-tRNA binding.
    C. mRNA binding.
    D. Binding to releasing factors.
    E. Amino acylation of tRNA.
19. Which of the following is TRUE of both eukaryotic and prokaryotic gene expression?
    A. A 3'-polyA tail and a 5-cap are added to mRNA.
    B. Translation of mRNA occurs while its transcription is in progress.

C. The mRNA is synthesized in the 5'⇒3' direction.
D. The transcription and translation are coupled.
E. Mature mRNA is polycistronic.

20. Why the genetic code is said to be degenerate?
A. Each codon name is formed of different bases.
B. The triplet codons differ from one amino acid to the other.
C. Many of the amino acids have more than one codon.
D. Wobbling in the $1^{st}$ anticodon base and the $3^{rd}$ codon binding.
E. Codons are used in an overlapping manner.

21. Several different proteins are synthesized from a typical prokaryotic mRNA because:
A. Several reading frames can be used.
B. The mRNAs contain several translation cistrons.
C. Alternative usage of several operators of the same gene.
D. Alternative splicing of the pre-mRNA (hnRNA).
E. Genes could be overlapping.

22. One of these enzymes is not a target for anti-cancer chemotherapy:
A. Dihydrofolate reductase.
B. Thymidine synthase.
C. Topoisomerases.
D. Telomerase.
E. Restriction endonucleases.

23. The expression of a gene could be obliterated at its mRNA level by:
A. Gene augmentation.
B. Gene rearrangement.
C. Small antisense oligonucleotide.
D. Ribosome.

24. For lac-Operon, absence of glucose causes:
A. A decrease in the cAMP level.
B. Inactivation of the CAPP.
C. Inactivation of the repressor protein.
D. A decrease in the transcription level of the lac-gene.

25. All of the following factors are required for the functionality of the lac-operon EXCEPT:
A. The PCR.
B. The permease.
C. The CAP sequence.
D. The operator sequence.

26. The messenger RNA (mRNA) is composed of all EXCEPT:
A. The telomeric sequence.
B. The cap 7-methy-GTP.
C. The start of translation sequence.
D. Sequences that controls its half-life through binding into regulatory proteins.

27. Considering transfer RNA (tRNA):
A. It is processed from a large transcript.
B. The initial transcript is intron-less.

C. The 3'-ACC sequence is acquired from the gene.
D. The D loop and arm is recognized by specific amino-acyl-tRNA synthetase.

28. Functions of the mRNA capping include:
    A. Destabilizing the mRNA and shorten its half-life.
    B. Helps transporting mRNA back into the nucleus.
    C. Controlling rate of mRNA translation.
    D. Enables production of more than one protein from the same mRNA.
29. All of the following are functions of the polyadenylate 3'-tailing of the mRNA EXCEPT:
    A. Differential production of tissue-specific protein isoforms.
    B. Determines the intracellular location of the mRNA and the protein.
    C. Enhances mRNA half-life.
    D. Destabilizing the mRNA and shorten its half-life.
30. The presence of intron within a gene helps:
    A. Separating the functional domains of the gene.
    B. Allows recombination between exons of different genes.
    C. Allows alternative splicing of exons of one gene.
    D. May contain gene insulator sequences.
31. Considering reverse transcriptase - all are CORRECT except;
    A. Is more sensitive to inhibition by dideoxy nucleotides than regular DNA polymerase.
    B. Locates to adenoviruses.
    C. Synthesizes dscDNA for mRNA.
    D. Is an important tool for recombinant DNA technologies.
32. The eukaryotic gene expression control sequences that works orientation and location-independently to control higher than the basal transcription level is;
    A. The promoter.
    B. The TATA box.
    C. The enhancer.
    D. The operator.
    E. The polyadenylation site.
33. Considering both of the eukaryotic and prokaryotic mRNA;
    A. They require addition of a 3'-polyA tail and a 5'-cap.
    B. Translation of mRNA can begin before transcription is complete.
    C. The mRNA is synthesized in the 3 to 5 direction.
    D. RNA polymerase binds at a promoter region upstream of the gene.
    E. Mature mRNA is always precisely co-linear to the gene from which it was transcribed.
34. The genetic code is said to be degenerate because of which of the following?
    A. The gene and its RNA product are no colinear.
    B. The genetic code is formed of triplet nucleotide sequences.
    C. Many of the amino acids have more than one triplet codons.
    D. Wobbling upon the anticodon-codon binding.
    E. The different codons of one amino acid have the same 1$^{st}$ and 3$^{rd}$ nucleotides.

35. Several different proteins can be synthesized from a typical prokaryotic mRNA because;
    A. The usage of the available reading frames.
    B. The usage of several terminator sequences.
    C. The usage of several operator sequences.
    D. Alternative RNA splicing.
    E. The mRNA is polycistronic.
36. In E. coli, under high lactose, high glucose conditions, which of the following could lead to maximal transcription activation of the lac operon?
    A. A mutation in the lac I gene (which encodes the repressor).
    B. A mutation in the CRP binding site leading to enhanced binding.
    C. A mutation in the operator sequence.
    D. A mutation leading to enhanced cAMP levels.
    E. A mutation leading to lower binding of repressor.
37. The reason there are 64 possible codons is which of the following?
    A. There are 64 aminoacyl tRNA synthetases.
    B. Each base is able to participate in wobbling.
    C. All possible reading frames can be used this way.
    D. There are four possible bases at each of three codon positions.
    E. The more codons, the faster protein synthesis can be accomplished.
38. Which of the following describes a common theme in the structure of DNA binding proteins?
    A. The presence of a specific domain that interacts with DNA bases across the major groove.
    B. The ability to recognize RNA molecules with the same sequence.
    C. Formation of hydrogen bonds between the peptide backbone and the DNA backbone.
    D. The presence of zinc.
    E. The ability to form dimers with disulfide linkages.
39. A mutation in the I (repressor) gene of a “non-inducible” strain of E. coli turned the lac operon off. Which of the following provides a rational explanation?
    A. The repressor has lost its affinity for inducer.
    B. The repressor has lost its affinity for operator.
    C. A trans acting factor can no longer bind to the promoter.
    D. The CAP protein is no longer made.
    E. Lactose feedback inhibition becomes constitutive.
40. The template 5'-GCTATGCATCGTGATCGAATTGCGT-3' DNA strand gives which of the following RNA sequences in what direction upon transcription?
    A. 5'-ACGCAATTCGATCACGATGCATAGC-3'.
    B. 5'-UGCGUUAAGCUAGUGCUACGUAUCG-3'.
    C. 5'-ACGCAAUUCGAUCACGAUGCAUAGC-3'.
    D. 5'-CGAUACGUAGCACUAGCUUAACGCA-3'.
    E. 5'-GCTATGCATCGTGATCGAATTGCGT-3'.

41. Differences between the reverse transcriptase and DNA polymerase include;
    A. Synthesizes DNA in the 5 to 3-direction.
    B. Contains 3 to 5-exonuclease activity.
    C. Follows Watson-Crick base pair rules.
    D. Synthesizes DNA in the 3 to 5-direction.
    E. Can insert inosine into a growing DNA chain.
42. The regulatory sequences in DNA are usually described in;
    A. Within introns.
    B. 5'-3'-sequences in the template strand upstream the gene.
    C. At 3'-end of the gene.
    D. 5'-3'-sequences in the coding strand upstream the gene.
    E. After the 1st exon.
43. The DNA binding activity of the catabolite activator protein is regulated by;
    A. Lactose.
    B. cGMP.
    C. Repressor protein.
    D. cAMP.
    E. Operon.
44. The eukaryotic DNA-dependent RNA polymerases are;
    A. One.
    B. Three.
    C. Four.
    D. Five.
    E. Two.
45. The eukaryotic mRNA is modified in the following ways;
    A. Glycosylation.
    B. Phosphorylation.
    C. 5'-capping and 3'-tailing.
    D. 5'-tailing and 3'-capping.
    E. Farnesylation.
46. Enhancers are unique gene expression control elements that;
    A. Decrease the rate of transcription.
    B. Work orientation and distance independently.
    C. Are tissue specific control sequences.
    D. Is a component of the basal transcription machinery.
    E. Contain TATA box.

*Answer key for the True/False questions:* 1, false; 2, true; 3, true; 4, true; 5, true; 6, true; 7, false; 8, false.

*Supply the Missing Information questions:* 1, specific functional RNA; 2, the gene proper/response elements/promoter/terminator; 3, transcription level; 4, sigma (σ)/rho (ρ); 5, small nuclear RNA/60 proteins; 6, post-maturation chemical modifications; 7, the gene proper/orientation/basal; 8, orientation/close/gene transcription; 9, cell-cell/cell -matrix /methylation/remodeling; 10, recombination/rearrangement/conversion/oncogenic /amplification; 11, termination-cleavage polyadenylation/intron-exon/exon/interfering/mRNA

/translation; 12, eIF-2; 13, eIF-4-binding protein; 14, initiation/misreading; 15, peptidyl transferase.

*Answer key for the MCQs*: 1, A; 2, A; 3, C; 4, A; 5, D; 6, E; 7, B; 8, C; 9, D; 10, C; 11, B; 12, C; 13, A; 14, D; 15, C; 16, B; 17, D; 18, E, 19, C; 20, C; 21, B; 22, E; 23, C; 24, C; 25, A; 26, A; 27, B; 28, C; 29, D; 30, D; 31, B; 32, C; 33, D; 34, C; 35, E; 36, D; 37, D; 38, A; 39, A; 40, C; 41, B; 42, D; 43, D; 44, B; 45, C; 46, B.

*Chapter IV*

# Applied Molecular Biology (DNA and Animal Cloning, Gene Therapy, and Stem Cell Biology)

## Topics Discussed

- DNA cloning and recombinant DNA technology
- Restriction endonucleases
- Cloning vectors
- Polymerase Chain Reaction
- Molecular analytical techniques for identifying DNA, RNA and protein
- Gene therapy
- Transgenic animals and stem cell biology
- Animal cloning
- Genomics, Transcriptomics, Proteomics, and Bioinformatics.

## Learning Objectives

After understanding this part the student should be able to:

- Explain the basis of the DNA cloning and recombinant DNA technology
- Differentiate between a cDNA clone and a genomic clone
- Describe PCR assay, pair of primers, product size and applications
- Describe the typical characteristics of restriction endonucleases with regard to their selection of cleavage sites on DNA and the types of fragment ends that are produced
- Describe the general steps used to perform a southern/northern/western blot and the type of information it provides

- Describe typical features of a bacterial plasmid as compared to other vectors used for cloning DNA
- Describe the general types and techniques used for gene therapy and methods of its transfer
- Differentiate between different types of transgenic animals and their uses
- Describe the basic foundations of animal cloning
- Differentiate between different types of stem cells and their applications.

Describe the basic foundations of the new sciences: genomics, transcriptomics, and proteomics, and, their integral usage in bioinformatics.

The advances in the study of genetic makeup of the organisms and the associated techniques over the last few decades has brought us to a stage where we can select a small gene in the plethora of genes in the chromosomes, isolate it, make copies and then insert it into other organisms for its expression. Today, complete DNA sequence of the genomes of a number of organisms (including humans) has been documented and now we are beginning to understand the function of more and more sequences of DNA.

# DNA Cloning

A *clone* literally means an identical copy or a replica. The term appears to have originated from microbial cultures where cells of a single type/organism grow isolated and reproduce to create a population of identical cells (colony). DNA cloning involves separating a DNA segment or a specific gene from a larger chromosome (genomic DNA cloning) and propagating it thousands or millions of times. A specific mRNA could also be cloned after its reverse transcription into dscDNA (cDNA cloning). The result is selective amplification of a particular gene or DNA segment. This cloning is used for many applications that include: DNA sequencing; studying the DNA structure-function relationship and regulation; production of recombinant proteins; gene therapy; and transgenic animal/plant production. Clones for overlapping pieces of the whole genome of an organism is called genomic library, whereas, clones carrying dscDNAs for all the RNA species of a cell is called cDNA library that is tissue- and developmental stage-specific. Genomic DNA libraries contain representative overlapping fragments of partially restriction digested genetic information (all genetic material; exons, introns, regulatory sequences, non-genic sequences, and expressed and non-expressed genes) of an individual organism and hence is organism-specific.

*Steps of DNA cloning* **(**see the figure for recombinant protein production**):**

a. *Precise cutting the target DNA:* Sequence-specific endonucleases (restriction endonucleases) act as molecular scissors to cut the identified DNA segment. Or using dscDNA for the gene's mRNA.
b. *Selecting a self-sustaining carriage and delivery agent, i.e., the vector.*
c. *Joining two DNA fragments:* The DNA ligase, e.g., T4 DNA ligase that uses ATP or E. coli DNA ligase that uses $NAD^+$ as energy source to joins the cloning vector DNA and target DNA (or dscDNA) utilizing ATP conversion into AMP + PPi for every

phosphoester bond formed. The resulting composite DNA comprising DNA from two or more sources are called recombinant DNAs.

d. *Inserting the recombinant DNA into a selected host cell:* The enzymatic machinery of the host cell is used for the multiplication of the inserted DNA and/or its expression into proteins.

e. *Confirmation of successful cloning* and *selection of the host cells that carry the inserted recombinant DNA*. Using specific radiolabeled DNA probe complementary to a sequence in the cloned DNA the positive cells could be distinguished and propagated. The DNA construct could be extracted for further characterization, e.g., DNA sequencing.

*Genetic engineering* is the technique that enables the molecular biologists to manipulate DNA of viruses, bacteria or eukaryotic cells *in vitro* or *ex vivo*. The DNA product of such manipulation is called recombinant DNA, the technique is popularly known as gene-splicing or genetic engineering, or recombinant DNA technology. Applications of the technology aim at studying the DNA structure-function relationships, improvement or modification of the genetic makeup of bacteria, plants or animals, production of recombinant human proteins, and treatment of genetic disorders.

## Components of the Recombinant DNA Technology

Along with the $1^{st}$ tool discussed before, that is *the reverse transcriptase*, recombinant DNA technologies requires; target DNA and/or cDNA from target mRNA, molecular scissors restriction endonucleases, cloning vectors, the amplifying polymerase chain reaction, the suitable cloning host cell (bacteria, yeast, and other eukaryotic cells), and DNA analytical techniques including DNA sequencing.

## Restriction Endonucleases

The restriction endonucleases are the $2^{nd}$ most important tool required for recombinant DNA technology first isolated in 1970 by Temin and Baltimore, and, comprise more than 100 types commercially available now. They are a group of remarkable bacterial enzymes that open the phosphoester bond in DNA at very specific sequences without removing the phosphate group, i.e., GATTCA into GAT-OH + P-TCA. Bacteria synthesize these enzymes as defensive mechanism against foreign DNA such as invading phages. Bacterial restriction endonucleases do not cleave their own genomic DNA due to the masking of their restriction sites by *Site-specific DNA methylases* that methylate the nucleotide residues at the restriction sites so as not to be accessible for cutting, thus protecting the native DNA from the action of restriction enzymes. Nowadays, the restriction map of the genome of several organisms including the human is known.

The restriction sites are a few base pair (4-10) long and have bilateral symmetry, i.e., the sequence is same on the two strands if read in 5'⇒3' direction. When restriction enzyme cleaves a DNA molecule, mostly the cut ends will be sticky (staggered). In this case, the complementary 'hanging' single stranded stretch of nucleotides of one double stranded DNA molecule bind the cut ends in another DNA molecule digested by same restriction enzyme. Thus, sticky ends can easily be used to combine more than one DNA fragments. The ends of the two DNA double strands in the recombinant molecule are then joined (linked) together by the DNA ligase.

The restriction enzymes are named after the bacterium from which they are isolated and take a number that indicates order of its isolation from that bacterium, e.g., HindIII, i.e., isolated from Haemophilus influenzae $R_d$ and the 3rd in order of discovery. Some restriction enzymes cleave DNA molecule leaving blunt ends such as HaeIII, i.e., isolated from Haemophilus aegyptius and the 3rd enzyme to be isolated from that bacterium.

**HindIII: 5'-A↓AGCTT-3'**
**3'-TTCGA↓A-5'**

**HaeIII: 5'-GG↓CC-3'**
**3'-CC↓GG-5'**

When the whole genome of one organism is restriction digested with one or several restriction endonucleases, the pattern of the produced fragments considering their length resolved by gel electrophoresis is specific to the species of the organism - with a few individual-specific difference in such pattern due to mutations (e.g., single nucleotide polymorphism) at some of the cut sequences. Such mutations lead to the cancellation of the cut sequence and fusion of two adjacent fragments. Fused fragments will show as a large fragment on the gel with the disappearance of the bands of the two corresponding normally expected fragments. Such restriction pattern is called *restriction fragment length polymorphism* (RFLP) that is species-specific and can reveal individual-specific pattern and/or disease causing mutations. The normal restriction pattern depends on the established restriction map of the genome of a specific species - where the position of sequences targeted by a specific restriction endonuclease is known; that is called *the restriction map*. For example, restriction digestion of the genome of the λ phage with HindIII gives very precise fragment size starting from 2 kb to largest of 23 kb - to the degree that they are used as molecular markers during electrophoresis.

## Cloning Vectors

The cloning vectors are the self-replicating DNAs (because of having an origin of replication) that use the host cell enzymes for its replication and transcription. They are the 3rd tool required for recombinant DNA technology in which a foreign piece of DNA could be inserted for the purpose of cloning. DNA fragments cloned into vectors can be isolated in large quantities. Vectors developed for variety of specific purposes; cloning DNA for sequence, structure and function studies, production of single-stranded DNA, gene therapy, transgenesis/knockout, or high level protein expression. Therefore, the choice of the vector depends on size of DNA molecules to be inserted and end use. The vectors are of several types as follows:

i. Plasmids: They are most convenient vehicles for bacterial recombinant DNA experiments. Plasmids are small extra-chromosomal circular double-stranded DNA found in bacteria and yeast. The plasmids are self-replicative and are composed of gene sequences transcribable into mRNA that direct synthesis of certain proteins. Naturally occurring bacterial plasmids range in size from 5 - 400 Kb. The only disadvantage of plasmids as DNA cloning vector is their limited capacity (6 - 10 Kb) and being unstable in eukaryotic cells. A typical plasmid is constructed from fragments from naturally occurring plasmids so as to contain; origin of replication sequence for independent replication (Ori), genes coding for enzymes that enable the host bacterium to resist the growth-inhibitory effects of certain antibiotics (e.g., ampicillin and tetracycline) as a selectable marker, and several recognition sequences for restriction endonucleases (seven are contained in the shown model pBR322 plasmid; Figure 59). The normal E. coli bacteria mostly used for cloning are sensitive to ampicillin and/or tetracycline, while the plasmid transformed bacteria would be resistant. This is why these antibiotic genes are called selectable markers in the cloning plasmid. Therefore, one of these antibiotic resistance genes will be a selection marker against contaminating wildtype E. coli (let say the ampicillin resistance gene), while the other tetracycline resistance gene will be utilized as recognition of recombination marker through insertional inactivation of its action. The bacterial colony with successful insert will be tetracycline sensitive while other colonies with the vector closed without the insert will be resistant.

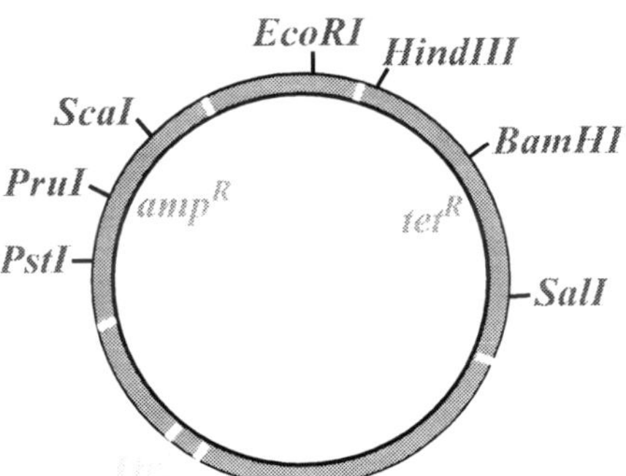


Figure 59. The pBR322 plasmid. The map shows the positions of the ampicillin-resistance gene (*amp* $^R$), the tetracycline-resistance gene (*tet* $^R$), the origin of replication (Ori) and the recognition sequences for seven restriction endonucleases (EcoRI, HindIII, BamHI, SalI, PstI, PruI and ScaI).

ii. Bacteriophages: They are bacterial viruses with linear DNA. About one-third of the phage genome is nonessential and can be replaced with foreign DNA. Bacteriophage lambda (λ) has a very efficient mechanism for delivering its 48,502 bp of DNA into a bacterium, and it can be used as a vector to clone somewhat larger DNA segments (see Figure 60). The maximum DNA that could be packaged into the infectious phage particles range from 40 – 53 kb long. Therefore, they can carry up to 20 Kb of foreign DNA. The cloned DNA is extracted from plaques of dead infected bacterial cells. Bacteriophage λ has a life cycle that can take two routes depending on the cellular status. Once inside the sensitive E. coli, the phage DNA circularizes utilizing its sticky cos ends. It may enter the lysogenic pathway where the phage DNA becomes integrated with the host genome and the phage DNA can be propagated with the host genome. This is suitable for recombinant protein production and DNA structure-function relationship studies. At some point a signal, e.g., stress signal/nutrient depletion triggers the lytic cycle of the phage DNA started with the release of its genetic material from the host genome, re-circularize and amplified, phage genes are expressed, a head is assembled and the phage DNA becomes packaged into the new heads, and a tail is assembled. When the level of mature phage particles in the cell reach a certain level, cell lysis occurs and mature phage particles are released. This would be suitable for DNA amplification for, e.g., sequencing and gene therapy.

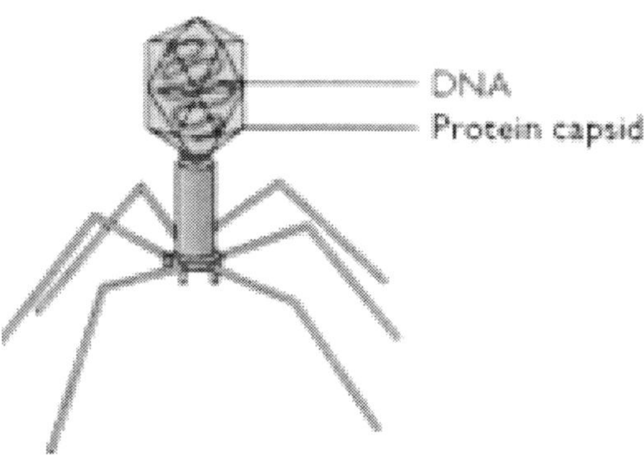


Figure 60. Bacteriophage λ complete particle.

iii. Cosmids: A typical cosmid is a plasmid that carries the λ phage sequence containing the cos site that is recognized by the packaging protein and enable the cosmid to be packaged into the λ phage particle. Particles containing cosmid DNA are as infective as real λ phages, but once inside the cell the cosmid does not direct synthesis of new phage particles and instead replicates as a plasmid. As any plasmid, cosmid has; an origin of replication (Ori); several recognition sequences for restriction endonucleases (two shown here); and, selectable markers (ampicillin-resistance in the model pJB8 cosmid shown; See, Figure 61). Because most of the phage genome was omitted, the cosmid can carry up to 50 Kb – the maximum phage particle space.

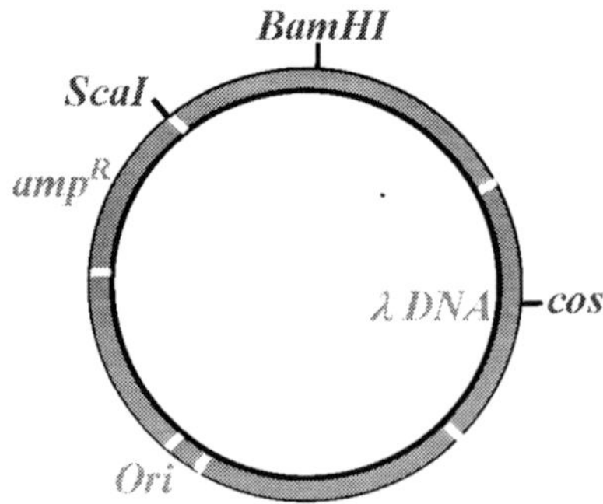


Figure 61. A model pJB8 cosmid. 5.4 kb in size and carries the ampicillin-resistance gene (*amp*$^R$), a segment of λ DNA containing the *cos* site, an *Escherichia coli* origin of replication (Ori), and the recognition sequences for restriction endonucleases (BamHI and ScaI).

iv. Bacterial Artificial Chromosomes (BACs): Bacterial artificial chromosomes are plasmids designed for the cloning of very long DNA segments (100 - 300 Kb). They generally include selectable markers such as resistance to the antibiotic chloramphenicol, as well as a very stable origin of replication (Ori). The large circular DNAs are then introduced into host bacteria by electroporation.
v. Yeast artificial chromosome (YAC): The genome of yeast, Saccharomyces cerevisiae, is very small and can be handled with ease particularly because the yeast is very easy to maintain and grow on a large scale in the laboratory. Furthermore, yeast provides an eukaryotic medium for cloning of the eukaryotic recombinant DNAs. The artificial circular chromosome (containing origin of replication, centromeric and telomeric sequences along with appropriate cloning and selection sequences) can carry ~3x106 bp DNA fragments. The source chromosome for the DNA is partially digested by appropriate restriction enzyme to separate the target DNA sequence. The YAC is opened by same enzyme to created complementary sticky ends. Then, the two molecules are subsequently sealed by the ligase and the circular YAC is opened linear with another restriction enzyme to create free telomeric ends and is subsequently transfected into yeast.
vi. Adenoviruses and retroviruses: The virus particles can be evacuated of their dangerous viral genes so as to be replication-defective and genes are replaced with the selected DNA for cloning. They can carry large amount of foreign DNA and deliver them, and particularly adenoviruses are the main for gene therapy.

## Polymerase Chain Reaction (PCR)

It is the $4^{th}$ important tool for recombinant DNA technology and other uses. Like DNA cloning, the PCR technique is used to amplify copy number of a DNA sequence and for the introduction of site-directed mutations. However, the PCR maximum fragment size to be amplified is around 400-600 bp that is much bigger with cloning. PCR has emerged as a very potent diagnostic tool particularly for the detection of the bacterial or viral infections.

A major step in PCR is to separate (denature) the DNA double strands by breaking down the hydrogen bonds without using the helicase. This is achieved in PCR by heating. However, at high temperature the DNA molecule could be damaged (e.g., strand breakage). Therefore several manipulations were employed to lower the required denaturation temperature. *Melting temperature, Tm,* is the temperature required to separate 50% of the bases in a given DNA helix. The higher the DNA cytosine and guanine content, the higher the Tm due to the three hydrogen bonds connecting CG, while AT connected by two hydrogen bonds only. The separation of the two strands can be followed by $UV_{260\ nm}$ absorbance that is almost doubled with the separation of the two strands. Re-association of the two strands requires lowering the temperature and increasing ion concentration, but depends also on DNA concentration and time. The denaturation and renaturation forms the very bases of polymerase chain reaction technology and is also important for nucleic acid hybridization techniques for identification, quantitation and separation of specific nucleic acid sequence. Tm is manipulated by:

i. *Changing the pH* that affects the ionization of bases and hence their binding.
ii. *Increase monovalent cation concentration* raises the melting point and *vice versa.*

iii. *Formamide and urea destabilize hydrogen bonds* and decrease Tm.

The bases of the cyclic three-stepped technology of thermal *P*olymerase *C*hain *R*eaction *(PCR)*, for synthesis of large amount of DNA using a few copies of template strands during gene manipulations are:

i. Melting DNA helix (i.e., separation of the two strands) by heating at ~90 - 96 °C or higher (*step 1, denaturation*).
ii. The reassociation of the two strands or a strand and a primer (*step 2, annealing*) at 50 - 65 °C or less.
iii. The DNA polymerization step synthesizes complementary strands at optimum temperature, 72 °C (*step 3, polymerization*), and so on; see Figure 62. This step uses the heat-resistant DNA Taq polymerase.

If the above three-stepped cycle is repeated 25 to 35 times so that every time the DNA strands from the previous cycle serve as the templates for subsequent cycle, several millions of copies of the initial very few copies of the original sequence would be produced within short time.

The specificity of the polymerization of the target DNA fragment is determined by the nucleotide sequence of the primers used. Two primers (known as forward and reverse primers) designed to base pair specifically at the 3'-ends of the two target DNA strands ensure the polymerization of the intervening DNA sequence only.

If the PCR is used to amplify an mRNA, the mRNA requires a prior conversion into a cDNA using *Re*verse *T*ranscriptase reaction. This kind of PCR is called accordingly *RT-PCR*. *Real-time PCR* is a version of PCR that allows the quantitative measurement of DNA or RNA molecules during their synthesis by following the rate of disappearance of a fluorescent component (the primers) due to its incorporation into the synthesized molecule.

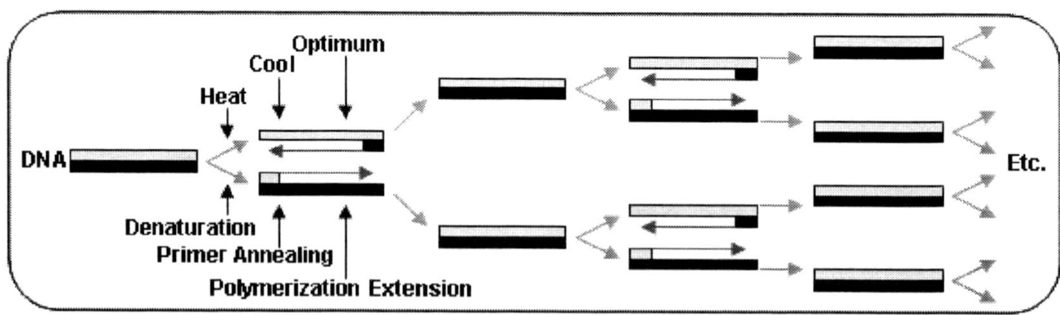


Figure 62. The three-stepped cyclic polymerase chain reaction.

*Worth Noting: Applications of PCR*

*PCR, RT-PCR and real-time PCR are used in:* diagnosis of infectious diseases, e.g., HIV, tuberculosis, schistosomiasis, and HCV; molecular cloning, DNA sequencing, molecular forensics, tissue typing for transplantation; detection of mutation in inherited and acquired genetic diseases; introduction of site-directed mutations; and in studying gene expression at the mRNA level quantitatively/qualitatively.

## *Transformation/transfection of the Host Cells by the Recombinant DNA-Vector*

- *Transformation:* The introduction of the recombinant vector into bacterial host cells is called transformation and the bacterium is called transformed bacterium. Transformation is conducted by using several methods including; self-delivery of phage vectors; electroporation, calcium phosphate or chloride, microinjection, and passive uptake and conjugation. Transformed bacteria bearing recombinant plasmids are induced to multiply to form a colony (a clone) and a large number of copies of that plasmid and hence of the inserted piece of DNA.
- *Transfection:* Transfection is the same as transformation but instead an eukaryotic host cell is used. Transfection is conducted by using several methods including; viral vectors, calcium phosphate or chloride, mechanical damage to cell membrane by scraping or sonication, microinjection and, liposomes reagents.

## **Some Molecular Analytical Techniques for Specific DNA, RNA, or Protein Identification** (Figure 63)

*Analysis of chromosomal DNA by Southern Blotting (after its inventor, E. Southern):* The technique aims at detecting a DNA strand depending on its specific sequence. The standard approach of DNA extraction include cell lysis by a suitable detergent, protein digestion to release the DNA by proteinase K, separation of DNA from tissue lipids and proteins by phenol-chloroform extraction and precipitation of DNA from the aqueous phase by ethanol and sodium acetate. The DNA precipitate is resuspended in aqueous buffer. The usage of a specific DNA probe with a complementary sequence helps detecting mutations that affects the hybridization of the probe and/or electrophoretic migration because – size wise - the smaller size molecules migrates the farther and faster most and vice versa. The basic steps of the technique include;

a. The extracted DNA is partially digested with restriction endonucleases.
b. The digest is separated into size- and shape-dependent bands by agarose gel electrophoresis.
c. The specific bands (including the target sequences band) are transferred onto a nitrocellulose or nylon membrane by blotting.
d. The specific sequence is identified by hybridization with radiolabeled (often $^{32}P$) or a chemical hapten labeled (such as biotin attached via a long side chain) probe. The probe is a specific complementary fragment of RNA, cDNA or genomic DNA.
e. The position of the bound label is visualized by autoradiography or by other means depending on the nature of the label after stringent washing to remove non-specific binding.

*Analysis of RNA by Northern Blotting:* The technique aims at specifically detecting RNA strand depending on its specific sequence. The rRNAs and tRNAs are of defined size and sequence and therefore relatively homogenous and can be purified by gel electrophoresis or density gradient ultracentrifugation. The mRNA is highly heterogeneous with respect to both size (~100 – 10,000 bases) and sequence except for the 3'-terminus, which in eukaryotic cells usually carries the poly-adenylic acid (poly A) tail. The poly(A) tail is usually long enough to allow mRNAs to be purified by affinity chromatography on oligo(deoxythymidine)-cellulose [oligo(dT)-cellulose]. RNA can be isolated as total cellular RNA or cytoplasmic and nuclear RNA if some fractionation of cells is carefully performed. Due to its inherent instability, the successful isolation of RNA depends on, use of strong lysis reagents, e.g., guanidium hydrochloride or guanidium isothiocyanate, the suppression of endogenous RNases during cell or tissue lysis, avoiding contamination with exogenous RNases during the isolation procedure, removal of contaminating DNA, protein, lipids and carbohydrates. The usage of a specific DNA/RNA probe with a complementary sequence helps detecting the level of expression of the mRNA and the nature of its isoforms in a specific tissue in one metabolic/developmental/disease status. Similar to DNA, size wise - the smaller size molecules migrates the farther and faster most and vice versa. The basic steps of the technique include:

a. Total RNA or total mRNA is extracted (by oligo-T coupled solid phase).
b. RNA is separated into size-dependent bands by gel electrophoresis in the presence of a denaturing agent to remove secondary structures.
c. The specific bands (including the target RNA band) are transferred onto nitrocellulose membrane by capillarity blotting.
d. The specific RNA is identified by hybridization with radiolabeled, fluorescently- or enzyme-labeled antibody probe of specific complementary fragment of cDNA or genomic DNA.
e. The position of the bound label is visualized by autoradiography or other suitable methods.

*Analysis of Protein by Western Blotting:* The technique aims at detecting a protein depending on its specific three-dimensional structure that affects its antigenic activity and affinity to bind specific antibodies. Like DNA and RNA electrophoresis, the protein resolution on a native gel depends on the size, net charge, three-dimensional structure and the electrophoretic environment. The usage of a specific poly- or mono-clonal antibody helps detecting the protein level and any antigenic changes due to, e.g., truncation of a specific antigenic epitope. The basic steps of the technique include:

a. Total protein or a specific compartmental protein is extracted.
b. The protein is separated into size- and charge-dependent bands by gel electrophoresis in a under native or denaturing conditions.
c. The specific bands (including the target protein band) are transferred onto a membrane by electric transfer in a wet or a semi-dry manner.
d. The target protein on the membrane is identified by specific binding with a radio-, fluorescently- or enzyme-labeled antibody as the probe.
e. The immune complex is visualized by autoradiography, fluorescent detection, or the specific enzyme activity on an appropriate substrate.

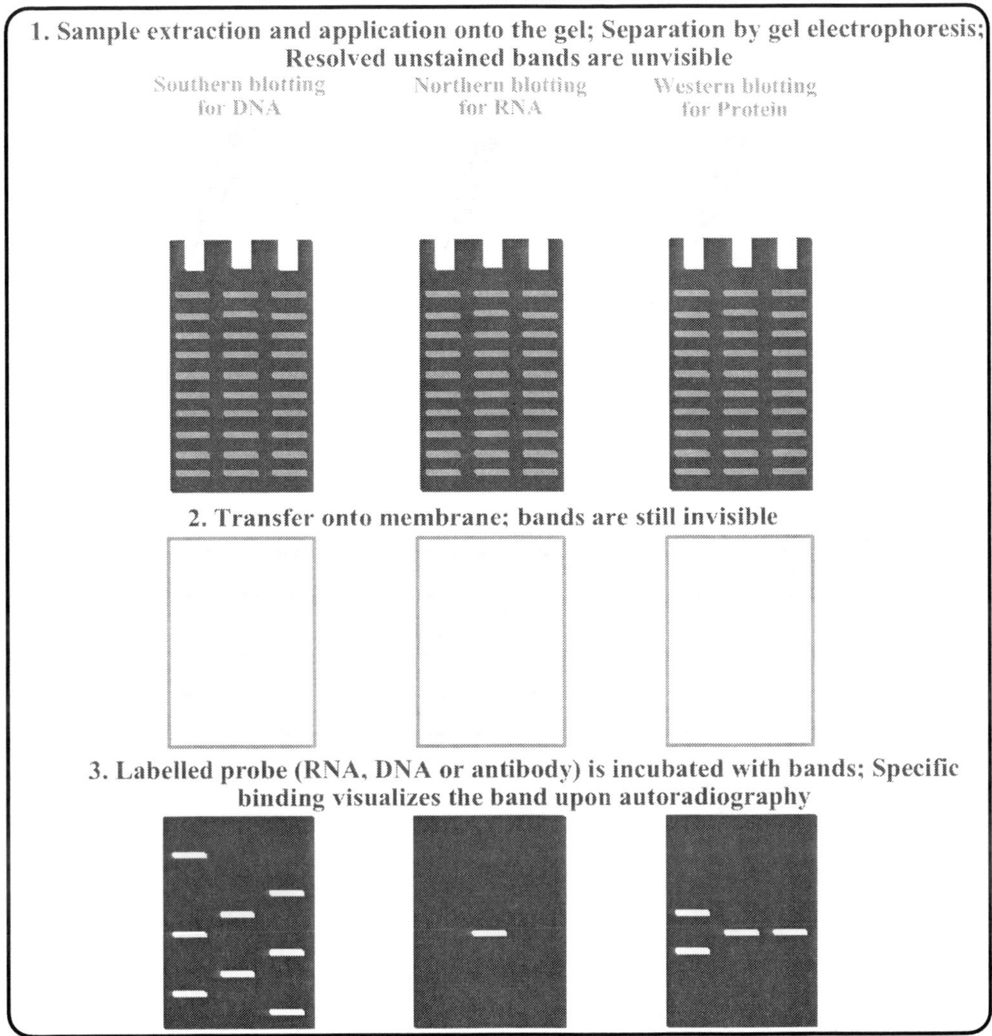


Figure 63. Identification of specific DNA, RNA or protein molecule by specific probe after resolution on the gel and transfer onto a special membrane to prevent gel interference.

## DNA Sequencing

Nowadays, the main method for sequencing nucleotides in a DNA fragment is the chain termination Sanger dideoxynucleotide method. Dideoxynucleotides are regular deoxynucleotides that also lack the -OH group on C3 of their 2-deoxyribose to be 2,3-dideoxy (ddATP, ddTTP, ddGTP, or ddCTP). The basic idea of the technique is that DNA polymerase works as usual to polymerize regular deoxynucleotides (dATP, dTTP, dGTP, or dCTP). However, in the presence of the dideoxynucleotide(s), random substrate competition allows the polymerase to incorporate the dideoxynucleotide in the growing DNA chain. This will terminate polymerization because the Dideoxynucleotides has no -OH on C3 of its deoxyribose. Knowing the size of the DNA fragment synthesized at termination - by resolution on sequencing gel - specify where the dideoxynucleotide was incorporated along the substrate DNA fragment.

If this process would be conducted in 4 separate reactions each containing only a specific dideoxynucleotide, the DNA polymerase reaction would stop due to competition between the deoxynucleotide (e.g., dATP) and its dideoxynucleotide form (ddATP) at all of the positions of the deoxynucleotide (i.e., dATP) due to incorporation of its dideoxynucleotide isomer (i.e., ddATP). This would create several DNA fragments of variable size dependent on where polymerization was stropped with the shortest fragment, the earliest and the closest to the 5'-end of the growing DNA strand because DNA polymerase polymerizes the chain in the 5'⇒3' direction.

Separating the different fragments produced in each reaction on a gel by electrophoresis that is able to resolve fragments differing by only one nucleotide in size, creates a pattern from the 5'-smallest (at the far most end of the gel; bottom) to the 3'-largest (at the near end of the gel; top, where reaction sample was applied). Resolving the products of the four reactions (A, G, C, or T) alongside each other in different lanes of the same gel allow reading the complementary sequence of the original DNA fragment by reading from the bottom to the top of the gel, i.e., going upwards from shortest in any lane to the subsequent in size in any lane in a zigzag manner towards the top last largest band; see Figure 64.

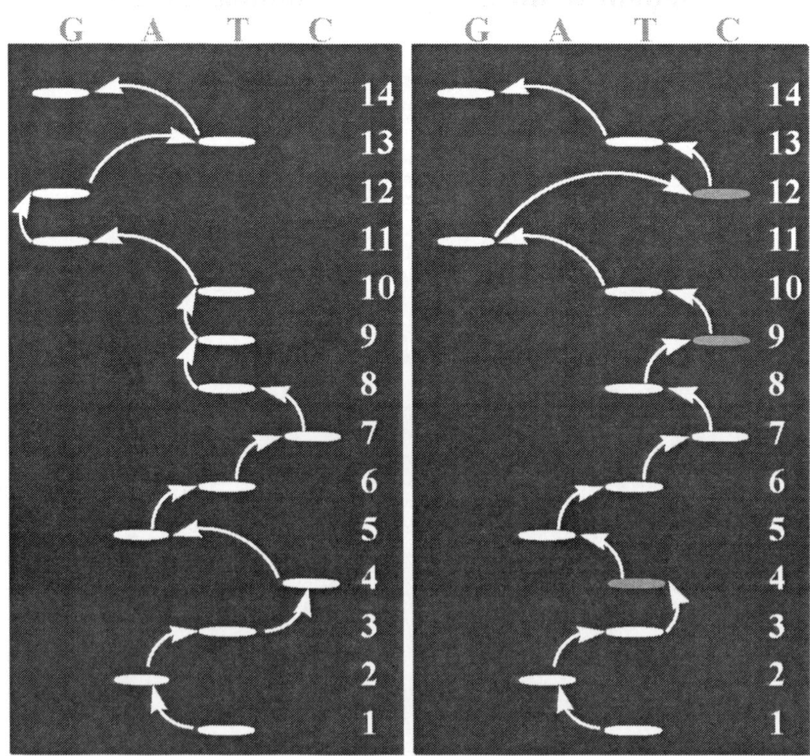


Figure 64. DNA sequencing gel of a normal and a mutant sample. The normal complementary sequence is 5'-TATCATCTTTGGTG-3' that was mutated at 3 nucleotides into 5'-TAT*T*ATCT*C*TG*C*TG-3'.

*Worth Noting: Other uses of the dideoxynucleotides*

Similar to their chain terminating effect on DNA polymerization, the dideoxynucleotides are used as chemotherapy particularly for the retroviral infection, e.g., HIV, and avian and swine influenza because the reverse transcriptase has a high affinity to the dideoxynucleotides than the cellular DNA polymerases.

## *Biomedical Applications of Recombinant DNA Technologies*

A. *DNA sequencing and gene mapping*. This was the prim aim of the human genome sequencing project. Cystic fibrosis (cystic fibrosis transmembrane conductance regulator, ATP-binding cassette (sub-family C, member 7; CFTR) is one of the most common genetically inherited disorders. Its gene was located to a 500kb region on chromosome 7q31.2 by DNA mapping and linkage procedures. The gene was one of the first human genes to which a genetic mutation was assigned by genomic DNA approaches. Cosmid and phage replacement genomic libraries were generated for this region of DNA. The clones were assembled and characterized by restriction enzyme analysis and by sequencing. Computer programs and manual comparisons were used to put in order the overlapping sequences at the ends of each clone - a process known a chromosome walking. The gene was identified as a 230kb stretch of DNA with 27 exons ranging from 38-724 bp. The gene sequence analysis helped understanding the pathogenetic mechanism of the disease mutations and developing treatment.

B. *Design of transgenic and gene-knock out* organisms, animals and plants: see below.

C. *Production of recombinant proteins for research, treatment and immunization as vaccine;* see Figure 65 for recombinant human proinsulin production as an example after cloning into bacteria. However, cloning eukaryotic nuclear genes into prokaryotes face several problems that should be addressed by engineering the bacterium first. These problems include; 1) differences in the epigenetic mechanisms; 2) absence of intron-removal system that could be circumvented by cloning dsDNA for the gene mRNA; 3) absence of the appropriate post-translation processing mechanisms, e.g.; proteases, and glycosyltransferases; 4) instability of the eukaryotic mRNA in the prokaryote; and, 5) instability of the protein product in the bacterium and lack of its secretion. Therefore, the most suitable for such cloning and protein production is the cDNA of the target gene. The mRNA is first isolated, converted into sscDNA then dscDNA, suitable restriction sequences are added to both ends, restriction digested, cloned into suitable vector, and transformed/ transfected into the suitable host cells.

D. *Molecular analysis of diseases and detection of mutations*. The sickle cell anemia mutant protein is detectable by Hb electrophoresis. It is also detectable by restriction fragment length polymorphism because the 5'-half of the gene contains a restriction site for the enzyme MstII (5'-CCTGAGG-3') in the β-globin gene that is abolishable by the mutation (5'-CCTG*T*GG-3'). Normal gene digestion produces a 1.1 kb size fragment, whereas, mutant gene will not be cleaved and produces 1.3 kb size fragment. Carrier case will have both fragments because one allele will be cleaved, whereas, the other will stay intact on Southern blot detection.

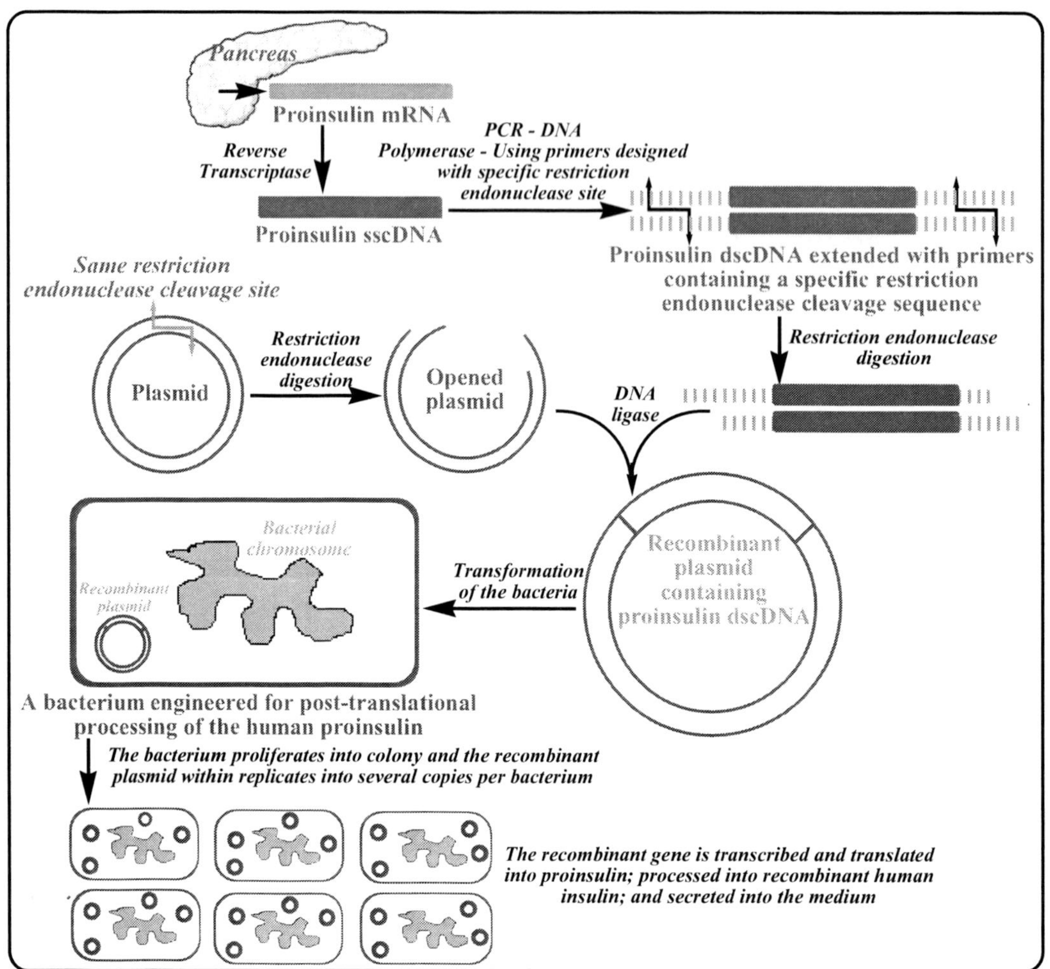


Figure 65. A simplified scenario of steps for gene cloning for recombinant protein production - insulin as an example.

E. *Pedigree analysis and forensic medicine*, e.g., paternity, identity and criminology (DNA fingerprinting). Human DNA fingerprinting target sequences are mainly the highly variable regions of microsatellite tandem repeats at certain loci in the genome. Their variation in size due to variation in the repeat number creates the length polymorphism among individuals. They are inherited and segregated in a strict Mendelian pattern, have a chromosome allele and individual-specific number of the repeat, and, have flanking individual-specific pattern of sites for restriction enzymes. Digestion with these restriction enzymes produces fragments that are individual-specific in their size on Southern blot; see Figure 66. Paternity and resolution of identity and criminal cases can be determined by this method. Only monozygotic twins will have identical patterns. If 13 sites match, the odds that any two individuals would possess such a fingerprint are so small - about $1:10^{12}$ - that the result can be considered a definitive match. The difference in their size (length polymorphism) can also be identified by amplifying them using PCR that utilizes primers targeting the repeat motif flanking sequences followed by resolving the products on agarose or denaturing polyacrylamide gel.

F. Gene therapy; see the details below.

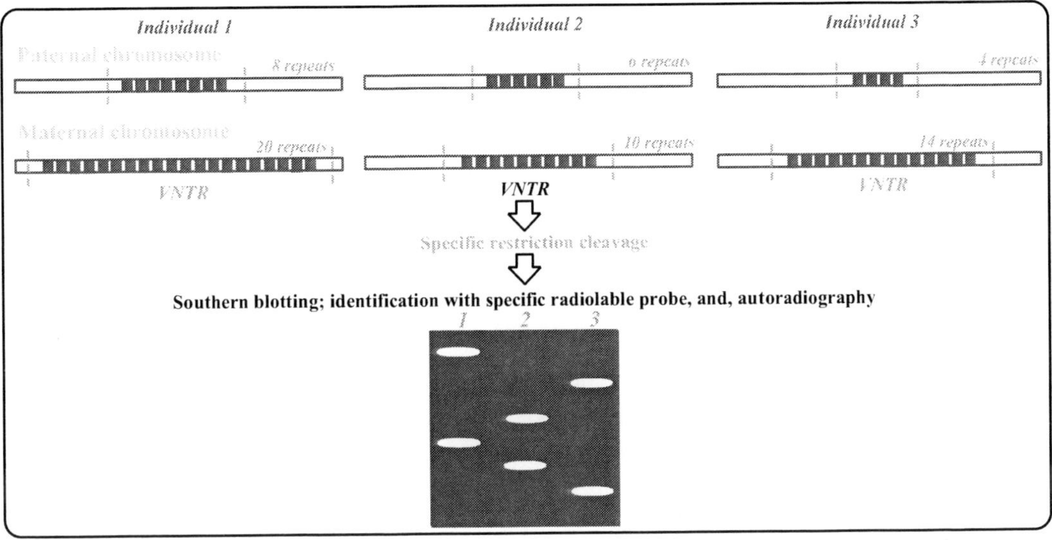


Figure 66. DNA fingerprinting using individual-specific differences in the variable number of tandem repeats (VNTR).

# Gene Therapy (Genotherapy)

Gene therapy involves the introduction (gene replacement) of an intact functioning gene to replace a defective gene or the destruction of an unwanted gene (gene disruption) so as to cure or control the disease process associated with the defective gene. It utilizes homologous recombination and specific viral integration mechanisms to integrate the replaced DNA sequence.

## The Gene Therapy vs. Traditional Approaches to Genetic Defects

The gene therapy techniques have shown limited success in actual practice. However, the techniques for gene therapy are improving day by day and we should see good therapeutic achievements in near future. Gene therapy is a radical and real full cure that would resolve the patient into completely productive individual as compared to the available traditional approaches (list below) for treating genetic diseases. There are some ethical issues associated with gene therapy in a number of countries. Generally, gene therapy is allowable for somatic cells only but not the germ cells.

*Available traditional approaches for treatment of the genetic diseases include:*

- Supplying deficient product such as thyroxin in goiter,
- Lowering unwanted product by limiting substrate availability such as diet low in phenylalanine in phenylketonuria,
- Replacing missing enzyme or protein such as insulin for diabetes, β-glucosidase for Gaucher's disease and blood clotting factor VIII for hemophilia,
- Inducing a deficient enzyme such as phenobarbital in Crigler-Najjar syndrome,
- Increasing activity of a mutant enzyme by large concentration of an activator such as vitamin B12 for methylmalonic aciduria, or
- Organ transplantation to supply missing protein such as blood transfusion for thalassemia, liver transplantation for galactosemia and myoblast transfer for Duchenne muscular dystrophy.

*The delivery of gene therapy is performed by two ways:*

1. *Direct transfer of genes into the patient systemically or topically* into certain tissue, e.g., onto the skin or as aerosol into respiratory system. Transfection is facilitated by packaging the non-viral recombinant vector into liposomes (spherical lipid-bilayer vesicles) or other biological microparticles. Alternately, the genes are packaged into genetically-engineered viruses, such as retroviruses or adenoviruses.
2. *By using engineered embryonic or adult stem cells in vitro* or *ex vivo*. This requires: isolation of the appropriate cells, their propagation in the laboratory, therapeutic gene introduction applying methods similar to those used in direct gene transfer, confirmation of the functionality of the gene, and reintroduction of the genetically-modified cells into the patient. Some cell types are able to localize to particular regions of the human body even if they were injected systemically, such as hematopoietic stem cells, which return to the bone marrow. This is called "homing" phenomenon.

## Applications of Gene Therapy

Gene therapy is developed primarily for treatment of inherited or acquired genetic diseases but it could also be used for improvement of human, plant and animal genetic characters (transgenic plants and animals), production of humanized animal organs and recombinant proteins.

Single gene inherited diseases such as immunodeficiency due to adenosine deaminase deficiency, cystic fibrosis; favism, sickle cell anemia, muscular dystrophy and thalassemia usually have a one defective gene leading to defective or deficient protein product and function. The feasibility of these diseases to be corrected by gene therapy exists.

Multi-gene complicated diseases such as cancer and complicated genome instability syndromes have several activated unwanted and/or inactivated beneficial genes due to mutations. Gene therapy for such diseases is complicated except if a major key regulatory gene could ameliorate the case, e.g., p53 for cancer - since restoration of the tumor suppressor function is by far more fundamental than correcting the oncogenic activities.

## Types of Gene Therapy

Gene therapy aims at manipulating the defective gene by either of two ways or specifically destroying the affected unwanted cell – in this case is cancer cell and saving the normal counterpart:

A. *Gene replacement, augmentation or correction:*
   a. *Gene correction*, i.e., correction of a mutation and restoring normal sequence and hence functions. It is technologically experimental today.
   b. *Gene augmentation (or amplification),* i.e., increasing the copy number of a defective gene that is giving a fraction of its normal activity may compensate into its normal functional level. It is only at experimental stages today.
   c. *Gene replacement* is reintroduction of a beneficial intact gene carried in a suitable vector into the target host cells to replace or to lie alongside the defective gene.

   Although the clinical application of the technique faces several problems (listed below), still the technology is medically very promising and has achieved real success *ex vivo* particularly in bone marrow transplants.

   Problems facing the realization of the clinical application of gene replacement therapy include:
   i. Absence of optimal vector (adenoviruses, retroviruses, and liposome-plasmid) that could deliver the gene into sufficient number of cells of the target organ or tissue.
   ii. Absence of precise way of organ, tissue or cell targeting.
   iii. Absence of accurate way of directed integration into a specific site of genomic DNA.
   iv. Absence of the normal physiological regulation of the recombinant gene.
   v. Repression of expression of the gene shortly after transfection (gene silencing).
   vi. There are deadly side-effects for the process with the available tools.

B. *Gene disruption (knockout) or reduction (knockdown):*

   Martin Evans and other colleagues were awarded the 2007 Noble Prize in Medicine for their pioneering work on stem cells and development of technologies leading to the generation of knockout mice. Nowadays, there are >10,000 knocked out genes in mice that give profoundly helped understanding genomics, genetics, and human disease, and helped developing new therapies. Gene knocked out is the reduction of the level of expression of a gene or the destruction of a damaging unwanted gene that is causing abnormal function in the cell such as; oncogenes that cause cancer, or genes that cause asthma or autoimmune diseases.

   1. *The expression of a gene could be obliterated at the DNA level by one of three ways:*
      i. Recombination with a non-specific foreign sequence (delivered by gene replacement approach) that disrupts and inactivates the gene.
      ii. Reverse orientation gene replacement. The mRNA produced from the original gene will be complementary to the mRNA of the replaced gene. They bind each other and prevent translation of one another (i.e., acting

as antisense RNA) and activated mRNA decapping and hydrolysis by RNaseH.

iii. Major groove-binding triplex-forming anti-DNA antisense oligodeoxynucleotides that prevents gene transcription by forming a triplex with the double stranded DNA. Triplex forming oligodeoxynucleotides bind to the duplex via Hoogsteen hydrogen bonding; e.g., T-A:T and $C^{+}$-G:C triplets that particularly target polypurine/polypyrimidine dsDNA.

2. *The expression of a gene could be obliterated at its mRNA level by one of three ways:*

   i. *Small antisense oligonucleotide* (~20 bp) that is DNA or RNA in nature with chemical modifications (phosphate, base and 2'-OH modifications) to extend their intracellular half-life against endogenous nucleases. More importantly, these modifications also make the oligonucleotide specifically complementary to a sequence in a specific mRNA - particularly flanking the translation start codon, AUG with enhanced recognition affinity. Oligonucleotide-based antisense techniques represent the most common and, to date, the most successful approach to achieving suppression or elimination of a genetic message. The oligonucleotide/mRNA hybrid sterically hinders translation in the ribosome and/or the hybrid is recognized by RNaseH that cleaves and hydrolyzes the mRNA strand. RNaseH is an endogenous enzyme that specifically cleaves the RNA moiety of an RNA:DNA or RNA:RNA duplex and is found in both the nucleus and the cytoplasm of all cells, where its normal function is to remove RNA primers from Okazaki fragments during DNA replication. Oligonucleotides as medicine need to be taken frequently like any other drug as they have an *in vivo* half-life of 1-3 days. The RNA oligonucleotides with 2'-O-alkyl protective modification cannot induce RNaseH cleavage of mRNA and only utilize ribosomal steric hindrance. This property was utilized to selectively suppress protein production from undesired splice variant of an mRNA, e.g., *in vitro* promotion of expression of wild type β-globin over the mutant β-globin variant in β-thalassemia. Moreover, antisense oligonucleotides that target intron-exon boundary could prevent intron excision from pre-mRNA were also used.

   ii. *Ribozymes*, they are specific enzymes that are RNA in nature and are natural or recombinant that are able to hydrolyze specific target mRNA so as to prevent its translation. Cells could be transfected with a ribozyme expression vector for long-term mRNA degradation.

   iii. *Small/short interfering RNAs (siRNAs):* This technology simulates the natural post-transcriptional gene silencing, where siRNAs are naturally produced endogenously in organisms including human. Establishment of double-stranded RNA (dsRNA) post-transcriptionally with a specific mRNA suppresses its expression, a process known as RNA interference (RNAi). The siRNAs are typically <400 nucleotides in length and could be synthesized *in vitro*, similar to the antisense oligonucleotides, or could be continuously expressed in the target cells. Intracellular transcription of siRNAs is achieved by cloning the siRNA templates into suitable vector(s). Cells transfected with a siRNA expression vector would experience steady, long-term mRNA inhibition, whereas, cells which are transfected with exogenous synthetic siRNAs typically recover from mRNA suppression within seven days or ten rounds of cell division.

C. *Destroying the defective cell (Replication-defective Oncolytic viruses):*

Adenovirus with its E1A and E1B oncoproteins suppresses the antitumor function of the tumor suppressor Rb and p53 proteins by protein-protein interaction. When an adenovirus enters a given host cell it has three main goals: to induce the cell cycle (to allow for the replication of virus genomes), to avoid premature death of the host cell (through cell suicide mechanisms or through the host immune system), and to exit the cell through lysis. The tumour suppressor protein p53 is central to both the control of the cell cycle and the control of programmed cell suicide (apoptosis), and thus represents a key target for adenovirus. A normal (wild-type) adenovirus encodes (at least) E1B 55kDa protein - among other proteins - that binds and inactivate p53 function. This stimulates cell cycle progression and prevents arrest the cell cycle and/or apoptosis. Consequently, the virus is replicated, packaged into particles and cell is lysed. In the recombinant oncolytic ONYX-015 form of the virus, the E1B 55kDa encoding gene is deleted, the virus can still infect a host cell and induce S phase, however, in normal cells with functional p53 and without E1B 55kDa, the cell either arrests its cycle or dies by apoptosis, both alternatives halt the virus life cycle and abort the infection process. In cancer cells with mutant inactive p53, ONYX-015 is able to infect, stimulates cell proliferation and completes a lytic infectious cycle. Unfortunately Onyx-015 has only shown limited clinical activity when combined with chemotherapies. Since then other virus-based therapies have shown greater promise for the treatment of cancer, most notably the oncolytic OncoVEX GM-CSF based on herpes simplex virus that expresses the immune stimulatory protein GM-CSF (Granulocyte-monocyte colony stimulating factor), which is now in Phase 3 clinical trials in melanoma following from Phase 2 studies where a remarkable level of efficacy in treating systemic disease was seen by stimulating a systemic anti-tumor immune response in the dying tumor micro-environment.

# Transgenic Animals and Animal Cloning

## Transgenesis

Transgenic animals (genetically-engineered animals) are those produced from embryos that incorporated cloned gene(s) or disrupted gene(s). Transgenic animals are thus a deliberate targeted modification of the genome - in contrast to spontaneous mutation. Foreign DNA is introduced into the animal using recombinant DNA technology through the germ line so that every cell, including germ cells, of the animal contains the same modified genetic material.

During the 1970s, the first *chimeric mice* were produced. The cells from early stage of development (eight cells) of two different embryos of different mice strains were combined together to form a single embryo that subsequently developed into a chimeric adult, exhibiting characteristics of both strains.

Today, transgenic animals using several animal species utilize three main approaches for gene transfer namely; *DNA microinjection* - first applied to mice in the early 1981s; *retrovirus-mediated gene transfer* and *embryonic stem cell-mediated gene transfer*. For their small size and low cost of husbandry in comparison to larger vertebrates, their short generation time, and their fairly well defined genetics, mice have become the main species used in the field of transgenics.

a) *DNA microinjection:* This method involves the direct microinjection of a chosen gene(s) construct from any genome source (allo- vs. xeno-), into the nucleus of a fertilized ovum. The introduced DNA may lead to the disruption; over- or under-expression of certain genes; or to the expression of genes entirely new to the transgenic animal. The manipulated fertilized ovum is transferred into the oviduct of a recipient female or uterus of a pseudopregnant foster mother (mated with a vasectomized male). A major advantage of this method is its applicability to a wide variety of species. Precautions are taken to optimize the gene incorporation into the recipient genome.
b) *Embryonic stem cell-mediated gene transfer:* The desired gene(s) are insertion into an *in vitro* cultured totipotent embryonic stem cell; or, the gene(s) are targeted for disruption through target mutagenesis using homologous recombination. The manipulated cell is then incorporated into an embryo at the blastocyst stage of development. The result is a chimeric animal; fortunately would have germ cells originating from the manipulated stem cell and transfer the modification to subsequent generations. Embryonic stem cell-mediated gene transfer is the method of choice for developmental gene knockout particularly in mice.
c) *Retrovirus-mediated gene transfer:* Due to their self-delivery, replication-defective retroviruses are commonly used as vectors to transfer genetic material into the cells. However, offspring derived from this method are chimeric, i.e., not all cells carry the retrovirus carried gene(s). Optimal transmission of the transgene occurs only if the retrovirus integrates into some of the germ cells.

*There is a very low success rate of getting first generation (F1) live birth animals containing the transgene. The gene requires confirmation for its expression level and regulation. Depending on the technique used, the F1 generation may result in chimeras. When the transgene has integrated into the germ cells, the so-called germ line chimeras are then inbred for 10 to 20 generations until homozygous transgenic animals are obtained and the transgene is present in every cell. At this stage, embryos carrying the transgene can be frozen and stored for subsequent implantation.*

### Transgenic Animals as Biotechnology

Transgenesis aims mainly at improving the genetic characteristics, biotechnological chemical production (protein products and humanizing animal tissues as potential organ donors), and/or studying gene function and regulation in the complex *in vivo* milieu and to study related human diseases.

Biotechnology utilizes *in vitro* cell fusion techniques; living organisms or their parts to make or modify products for commercial production, to change the characteristics of plants or animals, or to develop micro-organisms for specific uses. These genetic manipulations changed the long-held views as to what is considered to be animal, plant and human. Transgenic models are more precise in comparison to traditional animal models, for example the oncomouse with its increased susceptibility to tumor development enables results for carcinogenicity studies to be obtained within a shorter time-frame, thus reducing the course of tumor development in experimentally-affected animals. However, models are not strict normal system equivalents, so as their results are not directly extrapolatable to human usage.

*Worth Noting: Stem cell biology and applications*

Stem cells are undifferentiated unspecialized cells capable of indefinite self-renewal and are inducible to differentiation into other specialized cell types. The stem cells can theoretically divide indefinitely to replenish other cells throughout the life of the organism and senescence starts when the source of these stem cells diminishes.

Embryonic stem cells are at a very early developmental stage, and retain the flexibility to become any one of the more than 200 cell types that make up the human body. Such cells starting from the fertilized ovum (the zygote) and the cells of the inner mass of the early embryonic blastocyst (~30 cells) are known as *totipotent stem cells*. Adult or somatic stem cells that could be induced into a totipotent state include; hematopoietic stem cells, bone marrow stromal cells, breast and pancreatic ductal cells, oviduct cells, cumulus cells, basal epidermal cells, hepatocytes, and some non-neuronal brain cells.

Stem cells that differentiate into a large but limited number of cell types in a number of tissues or organs are called *multipotent stem cells*. Stem cells that differentiate into a number of cell types in one specific tissue or organ are called *pluripotent stem cells*. Stem cells that differentiate into one cell type are called *unipotent stem cells*.

The studying of the stem cell biology helps understanding the control of their differentiation into the dazzling array of specialized cells. This helps resolving diseases involving abnormal cell differentiation, e.g., cancer, teratogenesis and autoimmune diseases.

Self or embryonic stem cell lines grown *in vitro* could be genetically engineered to be used as spare *cell-based transplantation therapy* for; treating diseases (including; Parkinson's and Alzheimer's diseases, spinal cord injury, stroke, burns, heart disease, diabetes, osteoarthritis and rheumatoid arthritis) without facing the trouble of graft rejection as in the traditional transplantation. They are used also to overcome inherited or acquired genetic diseases of the individual.

## Animal Cloning

Cloning or artificial twinning of an existing organism (animal or human) is an asexual form of reproduction (vegetative form, like some plants) in which the creation of a cell, tissue, organ or an intact organism bypasses the conventional reproductive fusion of two half cells (the sperm and the ovum) by using a complete nucleus from a somatic cell and a nucleus-free ovum.

### Technique

The technology is known for long time at the level of simple biological systems like frogs. It got the big bang when applied to complicated animals, i.e., Dolly ewe in 1997 that opened the door for human cloning. In 1973, when a nucleus of a basal skin cell from an adult frog - that contains complete set of chromosomes and has a stem cell potential - was introduced into a frog ovum with its pronucleus previously destroyed by UV exposure, the cloned nucleus developed into an entire tadpole and cloned frog; see Figure 67.

### Animal Cloning through the Nuclear Transfer Technique Requires three Components

1) *Nuclear donor parent* (the male or the female) to donate a complete nucleus at G1 phase of cell cycle from a somatic cell that retains stem cell properties of self-renewal and a degree of pluripotency to differentiate into other cell types, e.g., breast and pancreatic ductal cells, oviduct cells, cumulus cells, and basal epidermal cells. These cells require to be starved in the *in vitro* culture to induce arrest of mitotic activity at G1 phase.

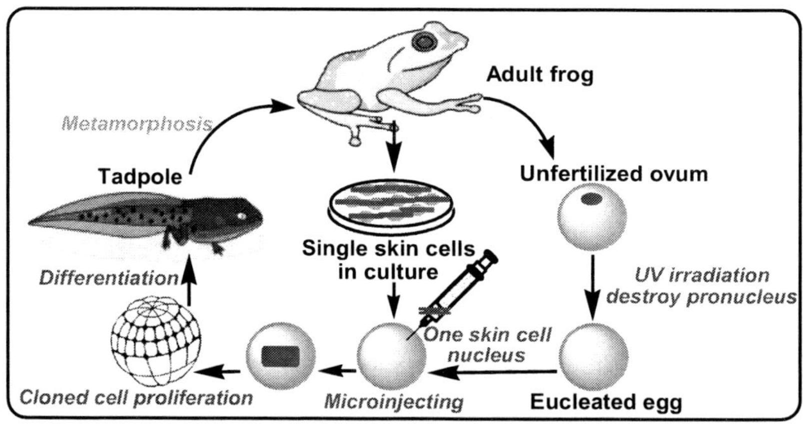


Figure 67. Steps used for the historic frog cloning.

2) *A nucleus-free (enucleated) late stage secondary oocyte ovum as a soil.* This exploits the fact that the ovum cytoplasm with its content of maternally stored mRNAs is the control point for the cloned or natural zygote division after introducing the complete nucleus up to the blastocyst stage. The transcription of zygotic genes starts late after the initial few cell cycles.
3) *Forster uterus* into which the cloned cell is transferred, implanted and grew into an identical organism. Example is the famous *Dolly* ewe. However, mammalian cloning efficiency is very low and may create defective embryos.

### *Disadvantages of Cloning*

i. Damage of the natural family life, diversity and balances among humanity.
ii. Possible creation of dangerous monstrous organisms, diseases and early aging.
iii. Creation of organ and tissue to the 3-dimensional normal form requires interaction of several cell types, relations to other tissues and normal foundation of nerve and blood supply, all are difficult to establish *in vitro* nowadays. However, single stem cells from cloned early and late embryos could nest into the appropriate tissues to repair cellular deficiencies. These cells could be used fresh or after chemical treatment with differentiating agents, e.g., retinoids, nicotinamide and transforming growth factor β (TGF-β).

*Worth Noting: Evidence for maternal mRNA control of early development:*

Early embryonic events are orchestrated through the posttranscriptional control of maternal mRNA. Fertilization acts as a trigger to initiate a program of events starting with cleavage, and continuing with blastula, gastrulation and neurulation, etc. Although fertilization results in union of maternal and paternal genomes, zygotic gene activity is not required until different levels of the zygote divisions that may reach the blastula stage. In fact, after fertilization and through cleavage, the maternal, paternal and zygotic genomes are completely dispensable. This is because, in the ovum (the secondary oocyte), there exists a stockpile of maternally derived mRNAs which govern embryogenesis through cleavage to the blastula stage. Following the formation of the blastula or earlier, zygotic gene transcription is activated, which carries the embryo through the rest of embryogenesis. *Evidence for maternal mRNA control of early development includes;*

1. *The dominance of the maternal trait in inter-specific hybrids:* Interspecies hybrids occur when sperm from a closely related species are used to fertilize an egg of another related species. It was evident that maternal contribution featured more prominently in early embryogenesis, where, maternal specific isozymes were found to be expressed exclusively in the early embryo until mid blastula after which paternal contribution was observed.
2. *Cleavage of the enucleate embryos following fertilization*, which was arrested at the blastula stage.
3. *Insensitivity of the cleavage to transcriptional inhibition* (with actinomycin D or α-amanitin) following fertilization. But it caused developmental arrest after cleavage at the time corresponding to when zygotic gene transcription is first detected.
4. *Prevention of the cleavage and zygote embryonic death due to translational inhibition* (with puromycin) following fertilization. This demonstrated that new protein had to be synthesized from maternal mRNA. This observation correlated well with the surge in translation which was observed immediately after fertilization and suggested that stockpiled maternal mRNA was the key material required for the early embryonic program. This program in turn was executed by a translational activation; i.e., fertilization per se, which results in 10- to 40-fold stimulation of translation. Also, a significant translational activation of maternal mRNAs occurs at oocyte maturation in preparation for the early very rapid steps of embryogenesis.

Since cleavage could occur without transcription or nucleus, maternal material other than DNA is directing early development, i.e., maternal mRNAs. In Drosophila, the maternal bicoid protein is a transcription factor that plays a key role in governing development of the anterior end of the embryo. In oocytes an anterior to posterior mRNA gradient of bicoid is formed. This gradient is translated into protein at egg deposition (fertilization). Other gene products that affect anterior development indirectly are exuperantia, swallow and staufen proteins through governing the positioning of bicoid mRNA in the developing oocyte. Exuperantia is required for the early stages of bicoid localization, swallow for the intermediate stages while staufen protein acts at the terminal stages just prior to fertilization. Clearly from this example, key regulatory proteins in oocyte development have to be present at different times. Therefore, while the bulk of translation may occur at maturation/fertilization, a program of maternal mRNA translation must occur in advance of "the big one". Since transcription is completely dispensable in early development, this suggests that most or all of the translational machinery is present in the oocyte at the time of maturation and/or fertilization. *Translational suppression of maternal mRNAs before fertilization may be regulated by;*

a. *Maternal mRNA modification* (e.g., polyA, cap methylation, hairpin structures at the 3'- and 5'-UTR).
b. *Masking by mRNA-binding regulatory proteins* at specific sequences at the 3'- and 5'-UTR.
c. *Sequestration of mRNA into maternal pronucleus until fertilization or into cytoskeletal proteins* where the translational machinery locates in the oocyte. This co-localization of specific mRNAs may enhance their chances of being translated. Oocyte factors that unwind mRNA secondary structure are more active post fertilization.
d. *Alterations of key translational regulatory protein.*

# Genomics, Transcriptomics, Proteomics, and Bioinformatics

## Genomics and the Human Genome Project

The human genome project officially began in 1990. The ultimate goal was to generate a high-quality reference DNA sequence for the 3 billion bp human genome and to identify all human genes. Other important goals included sequencing the genomes of model organisms to interpret human DNA by comparative genomics and studying human variation.

In June 2000, the completion of the first working draft of the entire human genome was announced. First analyses of the details appeared in the February 2001 issues of the journals *Nature* and *Science*. The high-quality reference sequence was completed in April 2003, marking the end of the Human Genome Project. Coincidentally, it also was the 50$^{th}$ anniversary of Watson and Crick's publication of DNA structure that launched the era of molecular biology.

The human genome reference sequence provides a magnificent and unprecedented biological resource that will serve throughout the century as a basis for research and discovery and, ultimately, practical applications. The sequence already is having an impact on finding genes associated with human disease. Beyond sequencing, growing areas of research focus on identifying important elements in the DNA sequence responsible for regulating cellular functions and providing the basis of human variation. Perhaps the most insistent challenge is to begin to understand how all the "parts" of a cell; genes, proteins, and many other molecules work together for the cell to come alive and to be integrated in rather more living organisms.

The genome size does not reflect the total number of the genes content and does not reflect the complexity of the biological system but rather reflects the presence or absence and amount of the genome-wide repetitive sequence, redundancy and gene-intervening sequence. While the human genome is 3.2 Gb it has 30,000-40,000 protein-coding gene, whereas, the simple 125 Mb genome of the earth worm, *Caenorhabditis elegans* or 97 Mb of the Vetch plant, *Arabidopsis thaliana*, contains 25,500 genes and 19,000 genes, respectively.

Such sequencing helped identifying genes implicated in >1200 genetic disorders. It also showed that most of the apparently sound humans carry a few defective genes - since such genes only contribute to disease susceptibility with weak genetic penterance to represent themselves, i.e., require accumulative defects in other genes. In the near future, DNA tests will help designing personalized therapy for many diseases, and predict future illness, criminal readiness and life expectancy.

*One could conclude the following facts from the human genomic DNA sequence:*

- The human genome contains 3.2 billion chemical nucleotide bp.
- The average gene consists of 3000 bp, but sizes vary greatly, with the largest known human gene being dystrophin at 2.4 million bp.
- Functions are unknown for more than 50% of discovered genes.
- The human genome sequence is almost exactly the same (99.9%) in all people.
- About 2% of the genome encodes instructions for the synthesis of proteins.
- Repeat sequences that do not code for proteins make up at least 50% of the human genome.
- Repeat sequences are thought to have no direct functions, but they shed light on chromosome structure and dynamics. Over time, these repeats reshape the genome by rearranging it, thereby creating entirely new genes or modifying and reshuffling existing genes.
- The human genome has a much greater portion of repeat sequences than other forms of life. Therefore, the genetic key to human complexity lies not in gene number but in how gene parts are used by alternative splicing and extensive post-translational chemical modifications of proteins and the repertoire of regulatory mechanisms controlling these processes.
- Over 40% of predicted human proteins share similarity with fruit-fly or worm proteins.
- Genes appear to be concentrated in random areas along the genome, with vast expanses of non-coding DNA between.

- Chromosome 1 (the largest human chromosome) has the most genes (2968), and the Y chromosome has the fewest (231).
- Genes have been pinpointed and particular sequences in those genes associated with numerous diseases and disorders including breast cancer, muscle disease, deafness, and blindness.
- Millions of locations where single-base DNA differences (single-base alteration or polymorphism) occur in humans. This information promises to revolutionize the processes of finding DNA sequences associated with such common diseases as cardiovascular disease, diabetes, arthritis, and cancers.

## The Transcriptome

Transcriptomics is the study of the nature and amount of the protein-coding mRNAs in a given normal or diseased cell. The transcriptome is the most significant component of the RNA content of a given cell - despite constituting □4% of that total content - because it contains the cell-specific protein-coding RNAs that specify the composition of the proteome and hence determine the structural and biochemical capacity of the cell. A given cell type inherits its specific transcriptome repertoire ready-made from the mother cell and daughter cell uses *de novo* transcription to maintain the homeostasis of such catalog throughout its life-span. Difference between mother and daughter cells is brought about when either faces condition that necessitates differential gene usage. Studying normal transcriptome of a cell or differential transcriptome between a normal cell and same cell in an abnormal condition or a disease status is very essential for understanding normal physiology and diseases at the cellular level. Therefore, a comprehensive description of the transcriptome of cells is the foundation of a complete understanding of the complexities of proteome, the regulation network, and disease phenotypes.

## The Proteome

The proteome is the constellation of all proteins in a cell as the final product of genome expression at a particular time in a particular condition. As the cell physiology gets complexed, it transcriptome and proteome repertoires get complexed, e.g., an active cell such as a hepatocyte contains 10,000–20,000 different proteins ($\sim 8\times 10^9$ individual molecules) that represents ~0.5 ng of protein or 18–20% of the total cell weight. The copy numbers of individual proteins is not fixed and could be as scarce as 20,000 copy molecules to as abundant as 100 million copy molecules per cell. Therefore, proteome (the number of different proteins) in mammals is much larger than the gene number because alternative splicing, RNA editing and posttranslational modifications generate protein variants from single genes. Proteomes of mammalian cell do not vary much considering the abundant proteins because most of them are every-cell-necessary repertoire of *housekeeping proteins* that carry general biochemical and structural activities. The differentially-expressed *cell-specific proteins* are those of the scarce copy number, however, exceptions include the very abundant hemoglobin in red blood cells. The cell proteome is its genome executive and maintainer with the transcriptome as an intermediate. Unlike the relatively unchanging genome from one cell to the other, the dynamic proteome changes from one cell to the other and from minute to minute in response to the ever changing intra- and extracellular environment in the same cell. Studies that explore the repertoire of the cell protein structure and activities are known as *proteomics*.

Genome » Transcriptome » Proteome » Biochemical activities (Metabolomics)

## Bioinformatics

Bioinformatics or computational biology is the term coined for the new field of science - the "omics" age - that merges molecular, cellular, and whole-organism biology, computer science, and information technology to manage (collect, store) and analyze the biological data (genomic, transcriptomic and proteomic data), with the ultimate goal of understanding and modeling living systems and generating new biological information. Applying computational approaches facilitates the understanding of various biological processes such as a more global perspective in experimental design and the ability to take advantage of database mining to generate testable hypotheses regarding the function or structure of a gene or protein of interest by identifying similar sequences in better characterized organisms. It is mainly used to analyze DNA, RNA, and protein sequence homology and functions.

## Review Questions

A. *Write short notes on:*
   1. The basic requirement of animal and human cloning.
   2. Restriction endonucleases.
   3. Southern blotting for DNA.
   4. Gene transfer therapy.
   5. Different applications of recombinant DNA technologies.
   6. Stem cell characteristics and uses.
   7. Proteomics.

B. *Explain each of the following:*
   1. Differences between phage, plasmid and cosmid as experimental vectors for DNA cloning.
   2. The PCR technique and uses.
   3. Steps utilized for DNA cloning.
   4. Gene knockout therapy.
   5. The different techniques for gene transfer for generation of transgenic animals and their uses.
   6. The basic technique to DNA fingerprinting.
   7. The different uses of the dideoxynucleotides.
   8. Basic technique for DNA sequencing.

C. *True and False Questions:*
   1. The gene replacement aims at destroying an unwanted gene.

2. The gene knockout replaces a functional copy of a gene for its unfunctional copy.

D. *Supply the Missing Information Questions:*

1. The basic three steps of the Polymerase Chain Reaction (PCR) include.
    a. Separation of the DNA --------------------.
    b. The ---------------- of the two strands or a strand and a primer.
    c. DNA ----------------------.
2. Applications of PCR include:
    b. Diagnosis of ------------------ diseases.
    c. Tissue typing for ------------------------.
    d. Detection of disease-related ----------------------.
    e. Introduction of ------------------- mutations.
    f. Studying gene expression at the -------------- level.
3. The delivery of gene therapy is performed by two ways:
    **a.** Direct transfer of ------------ into the patient systemically or topically into certain tissue.
    **b.** Indirect engineering of embryonic or adult ----------------.

E. *Multiple choice questions:*

1. DNA can be denatured by:
    A. Heat.
    B. Changing the concentration of monovalent cations.
    C. Changing the pH.
    D. Formamide.
    E. All of the above.
    F. None of the above.
2. All are required for the PCR technique, EXCEPT:
    A. The reverse transcriptase.
    B. Nucleotide triphosphates.
    C. Primase enzyme.
    D. Specific primers.
    E. Heat-resistant DNA polymerase.
3. Applications of the PCR technique include all, EXCEPT:
    A. Diagnosis of infectious diseases.
    B. Genetic tissue typing for transplantation.
    C. Detection and introduction of mutation.
    D. Studying gene expression at the mRNA level.
    E. Animal cloning.
4. Considering restriction endonucleases:
    A. Are bacterial enzyme.
    B. Convert RNA into DNA.
    C. Hydrolyze proteins into single amino acids.
    D. Their mutation cause DNA repair related diseases.
    E. Are one part of the human immune system.
5. Is an important tool for recombinant protein production:
    A. Peroxisome.

B. Reverse transcriptase.
C. RNA polymerase.
D. Host organism.
E. A+C.
F. B+D.

6. A Typical cloning plasmid is having all, EXCEPT:
   A. Origin of replication.
   B. Recognition sequences for restriction endonucleases.
   C. Antibiotic selectable markers.
   D. A binding sequence for Sigma factor.
   E. A double stranded circular DNA structure.
7. All are required for recombinant protein synthesis EXCEPT:
   A. Plasmid.
   B. A non-specific DNA sequence.
   C. Restriction endonucleases.
   D. RNA polymerase.
   E. Reverse transcriptase.
8. Animal cloning requires:
   A. Nuclear donor cell.
   B. Restriction endonuclease.
   C. A suitable cloning vector.
   D. PCR.
   E. Mitogenic factors.
9. Considering the restriction endonucleases:
   A. They are eukaryotic nucleases.
   B. They have a specific cut sequences.
   C. They have DNA polymerase activity.
   D. They help converting the retroviral genetic RNA into dscDNA.
   E. They have proofreading activity.
10. Considering the restriction endonucleases:
    A. They are isolated from bacteria.
    B. The human genome does not have specific cut sequences for them.
    C. They are named using the regular systematic naming of enzymes.
    D. They are ribozymes.
    E. They remove DNA supercoiling during replication.
11. Ribozymes;
    A. Are RNA molecules with enzymatic activity.
    B. Are used in gene therapy.
    C. Are included in some of the spliceosome activities.
    D. Ribosome includes some such activity.
    E. A + C.
    F. B + D.
    G. All are correct.
12. Which of the following is responsible for retaining the enzyme protein disulfide isomerase in the ER?
    A. Glycosylation.

B. The presence of lys-asp-glu-leu sequence at the C-terminal.
C. Retention of the signal peptide.
D. Phosphorylation of Ser 22.
E. Attachment of a fatty acid.

13. Electrophoresis resolves double-stranded DNA fragments based on which of the following?
A. Sequence.
B. Molecular weight.
C. Isoelectric point.
D. Frequency of CTG repeats.
E. Secondary structure.

14. Which of the following sets of reagents are required for dideoxy chain DNA synthesis in the Sanger technique for DNA sequencing?
A. Deoxyribonucleotides, Taq polymerase, DNA primer.
B. Dideoxyribonucleotides, deoxyribonucleotides, template DNA.
C. Dideoxyribonucleotides, DNA primer, reverse transcriptase.
D. Two DNA primers, template DNA, Taq polymerase.
E. mRNA, dideoxynucleotides, reverse transcriptase.

15. Upon electrophoretic resolution of a mixture of DNA fragments of different size and shape, all are correct EXCEPT;
A. The larger DNA fragments migrate farther in the gel.
B. DNA fragments migrate toward the negative charge (anode).
C. DNA can be visualized using UV light and the dye ethidium bromide.
D. Total human genomic DNA cut by a specific restriction endonuclease will generate distinctly separable bands.
E. DNA must be denatured before it can be run in the gel.

16. In a pancreatic ductal adenocarcinoma model, the transcriptional expression of carbonic anhydrase as a differentiation marker is best determined by;
A. Genomic library screening.
B. Genomic Southern blot.
C. Tissue Northern blot.
D. Tissue Western blot.
E. VNTR analysis.

17. Which of the following primers would allow copying of the single-stranded DNA sequence 5' ATGCCTAGGTC?
A. 5' ATGCC.
B. 5' TACGG.
C. 5' CTGGA.
D. *5' GACCT.*
E. 5' GGCAT.

18. Which of the following tools of recombinant DNA technology is INCORRECTLY paired with one of its uses?
A. Restriction endonuclease - production of DNA fragments for gene cloning.
B. *DNA ligase - enzyme that cuts DNA, creating sticky ends.*
C. DNA polymerase - copies DNA sequences in the polymerase chain reaction.
D. Reverse transcriptase - production of cDNA from mRNA.

E. Electrophoresis - RLFP analysis.

19. The "Southern" technique involves:
    A. The detection of RNA fragments on membranes by specific radioactive antibodies.
    B. *The detection of DNA fragments on membranes by a radioactive DNA probe.*
    C. The detection of proteins on membranes using a radioactive DNA probe.
    D. The detection of proteins on membranes using specific radioactive antibodies.
    E. The detection of DNA fragments on membranes by specific radioactive antibodies.
20. DNA from a eukaryotic organism is digested with a restriction endonuclease and the resulting fragments cloned into a plasmid vector. Bacteria transformed by these plasmids collectively contain all of the genes of the organism. This culture of bacteria is refereed to as a:
    A. Restriction map.
    B. RFLP profile.
    C. F' factor.
    D. *Library.*
    E. Lysogenic phage.
21. Which of the following is not part of the normal process of cloning recombinant DNA in bacteria?
    A. Restriction endonuclease digestion of cellular and plasmid DNAs.
    B. Production of recombinant DNA using DNA ligase and a mixture of digested cellular and plasmid DNAs.
    C. *Separation of recombinant DNAs by electrophoresis using the Southern technique to determine where the desired recombinant migrates.*
    D. Transformation of bacteria by the recombinant DNA plasmids and selection using ampicillin.
    E. Probing blots of bacteria clones with radioactive DNA complementary to the desired gene.
22. Which of the following is not part of the normal process of cloning recombinant DNA in bacteria?
    A. Restriction endonuclease digestion of cellular and plasmid DNAs.
    B. Production of recombinant DNA using DNA ligase and a mixture of digested cellular and plasmid DNAs.
    C. *Separation of recombinant DNAs by electrophoresis using the Southern technique to determine where the desired recombinant migrates.*
    D. Transformation of bacteria by the recombinant DNA plasmids and selection using ampicillin.
    E. Probing blots of bacteria clones with radioactive DNA complementary to the desired gene.
23. Restriction endonuclease generated DNA fragments separated by gel electrophoresis and blot transferred onto a membrane filter are probed with a radioactive DNA fragment. This procedure is called:
    A. Gene cloning.

B. *The Southern technique.*
C. The polymerase chain reaction.
D. Recombinant DNA.
E. Gene mapping.

24. One of the most significant discoveries which allowed the development of recombinant DNA technology was:
    A. The discovery of antibiotics used for selecting transformed bacteria.
    B. *The identification and isolation of restriction endonucleases permitting specific DNA cutting.*
    C. The discovery of DNA and RNA polymerase allowing workers to synthesize any DNA sequence.
    D. The development of the polymerase chain reaction.
    E. The Southern technique for separation and identification of DNA sequences.
25. A key feature of insertional mutagenesis for the identification of plasmids containing recombinant DNA is:
    A. The production of nutritional auxotrophs.
    B. The DNA sequencing of recombinant plasmids.
    C. The production of restriction endonuclease maps of recombinant plasmids.
    D. Introns can be moved to new locations within the gene.
    E. *The disruption of a gene on the plasmid by the inserted recombinant DNA.*

*Answer key for the True and False Questions:* 1, true; 2, true.

*Answer key for the Supply the Missing Information Questions:* 1, double helix/reassociation/polymerization; 2, infectious/transplantation/mutations/site-directed/ mRNA; 4, genes/ stem cells.

*Answer key for the MCQs:* 1, E; 2, C; 3, E; 4, A; 5, F; 6, D; 7, D; 8, A; 9, B; 10, A; 11, G; 12, b; 13, E; 14, B; 15, A; 16, C; 17, D; 18, B; 19, B; 20, D; 21, C; 22, C; 23, B; 24, B; 25, E.

# Conclusion

In a very simple manner and rationale progress, the book highlighted the fundamental basics of medical molecular biology - the least required for understanding this area essential for all medical, veterinary medical, pharmacy, nursing, dentistry, genetics, biology, and biotechnology students. In every level it relates information to applied examples. This book hoped to establish a proper understanding for molecular biology within the medical and biology students' arena that is particularly important in this era of medical, pharmaceutical, and biotechnological applications of molecular biology.

# References and Further Reading and Web-Based Resources

## References and Further Reading Resources

Advances in Experimantal Medicine and Biology: Transgenesis and the Management of Vector-Borne Disease, Aksoy, S, editor, 2008. *Landes Bioscience and Springer Science + Business Media,* LLC. Landes Bioscience, Austin, Texas, USA.

At the Cutting Edge siRNA technology. Schütze N. *Molecular and Cellular Endocrinology,* 2004, 213, 115–119.

*Biochemistry and Molecular Biology Compendium,* Lundblad RL, editor, 2007. CRC Press, FL, USA.

Bioinformatics - *A Practical Approach*, Ye SQ, editor, 2008. CRC Press, Taylor & Francis Group, London, UK.

Cancer Biology: *An Updated Global Overview*, El-Metwally, TH, editor, 2009. *Nova Science Publishers, I*nc., Hauppauge, NY; 2009.

Coiled Spring*: How life begins,* Bier, E, editor, 2000. Cold Spring Harbor Laboratory Press, NY, USA.

Condensed Protocols from Molecular Cloning*: A Laboratory Manual,* Sambrook J and Russell D, editors, 2006. Cold Spring Harbor Laboratory Press, NY, USA.

*Discovering the Double Helix,* Watson J, editor, 2001. Cold Spring Harbor Laboratory Press, NY, USA.

*DNA From The Beginning*, Micklos D, et al., editors, 2002. Cold Spring Harbor Laboratory Press, NY, USA.

DNA mismatch repair and cancer. Li GM. *Front Biosci* 2003;8:d997-1017.

*Essentials of Molecular Biology*, Malacinski GM, editor, 4$^{th}$ edition, 2002. Jones and Bartlett Publishers, Sudbury, MA, USA.

*Gene Cloning: Principles and Applications.* Lodge J, et al., editors, 2007. Taylor and Francis, NY, USA.

*Gene Regulation, Latchman* DS, editor, 5$^{th}$ edition, 2005. Taylor and Francis, NY, USA.

*Gene Therapy Technologies, Applications and Regulations*, Meager A, editor, 1999. John Wiley & Sons Ltd, Hoboken, NJ, USA.

*Genes VII*, Lewin B, editor, 2000. Oxford University, Oxford, UK.

*Genomes 2,* Brown TA, editor, 2nd edition, 2002. BIOS Scientific Publishers Ltd, Oxford , UK.

*Handbook of Molecular and Cellular Methods in Biology and Medicine,* 2nd ed. Cseke L, et al., editors, 2004. CRC Press, Boca Raton FL, USA.

*Integrative Approaches to Molecular Biology,* Collado-Vides J, Magasanik B and Smith TF, editors, 1996. The MIT Press, London, UK.

Latest Developments in Gene Transfer Technology: Achievements, Perspectives, and Controversies over Therapeutic Applications. Romano, G, Micheli, P, PACILIO, C, Giordano, A. *Stem Cells* 2000; 18: 19-39.

Manipulating the Mouse Embryo: A Laboratory Manual, Nagy A, et al., , 3rd edition, 2003. Cold Spring Harbor Laboratory Press, NY, USA.

*Methods in Molecular Biology™ – 220: Cancer cytogenetics, Methods and proteocols, Swansbury J, editors, 2003. Humana Press, NJ, USA.*

*Methods in Molecular Biology™ – 224: Functional Genomics,* Brownstein MJ and Khodursky AB, 2003. Humana Press, NJ, USA.

*Methods in Molecular Biology™ – 226: PCR Protocols,* 2nd Edition, Bartlett JMS and Stirling D, 2003. Humana Press, NJ, USA.

*Methods in Molecular Biology™ – 263: Flow Cytometry Protocols,* 2nd Edition, Hawley TS and Hawley RG, editors, 2004. Humana Press, NJ, USA.

*Methods in Molecular Biology™ - 296: Cell Cycle Molecules and Mechanisms of the Budding and Fission Yeasts*, Humphrey T and Pearce A, editors, 2005. Humana Press, NJ, USA.

*Methods in Molecular Biology™ – 314: DNA Repair: Mammalian Systems,* Henderson DS, editor, 2nd edition, 2006. Humana Press, NJ, USA.

*Methods in Molecular Biology™ - 316: Bioinformatics and Drug Discovery,* Larson RS, editor, 2006. Humana Press, NJ, USA.

*Methods in Molecular Biology™ – 338: Gene Mapping Discovery and Expression, Methods and proteocols,* Bina M, editor, 2006.

*Methods in Molecular Biology™ – 556: Microarray Analysis of the Physical Genome Methods and Protocols*, Pollack JR, editor, 2009.

*MicroRNAs: Biology, Function and Expression,* Clarke N, et al., 2007. DNA Press, NY, USA.

*Modern Genetic analaysis*, Griffiths AJF, Gelbart WM, Lewontin RC, and Miller JF, editors, 2nd edition, 2002. W.H. Freeman and company, NY, USA.

*Molecular Biology in Cellular Pathology,* Crocker J and Murray PG, editors, 2003. John Wiley & Sons, Ltd, West Sussex, UK.

*Molecular Biology in Medicinal Chemistry,* Dingermann Th, Steinhilber D and Folkers G, editors, 2004. WILEY-VCH Verlag GmbH & Co. KGaA, Weinheim, Germany.

*Molecular Biology Made Simple and Fun,* Clark D and Russell L, 3rd edition, 2005. Cache River Press, Clearwater, FL, USA.

*Molecular Biology of Human Cancers - An Advanced Student's Textbook,* Schulz WA, 2005. Springer Science, NY, USA.

*Molecular Biology of the Gene,* Watson JD, et al., editors, 5th edition, 2004. Cold Spring Harbor Laboratory Press, NY, USA.

*Molecular Biology of the Neuron (Molecular and Cellular Neurobiology Series)* – 2nd edition, Davies RW and Morris BJ, 2004. Oxford University Press, NY, USA.

*Molecular Biology Problem Solver:* A Laboratory Guide, Gerstein AS, editor. 2001. Wiley-Liss, Hoboken, NJ, USA.

*Molecular Biology,* Clark D, editor, 2005. Elsevier Academic Press, San Diego, CL, USA.

*Molecular Cell Biology,* H. Lodish H, et al., editors, 5th edition, 2003. WH Freeman, NY, USA.

*Molecular Cloning: A Laboratory Manual,* Sambrook J and Russell D, editors, 3rd edition, 2001. Cold Spring Harbor Laboratory Press, NY, USA.

*Molecular Diagnostic PCR Handbook,* Viljoen GJ, Nel LH, and Crowther JR, editors, 2005. Springer, Dordrecht, The Netherlands.

*Molecular Mechanisms of Cancer,* Weber GF, editor, 2007. Springer, Dordrecht, The Netherlands.

*Molecular structure of nucleic acids: A structure for deoxyribose nucleic acid.* Watson JD, Crick FHC. Nature 1953;171:737-8.

*mRNA Processing and Metabolism,* Schoenberg R, editor, 2004. Humana Press, NJ, USA.

*Novartis Foundation Symposium 287: Mitochondrial Biology: New Perspectievs,* 2007. John Wiley & Sons Inc., Hoboken, NJ, USA.

*Recombinant DNA,* Watson JD, Gilman M, Witkowski J, Zoller M, editors, 2nd ed, 1992. Scientific American Books, WH Freeman, NY, USA.

*Retinoblastoma protein: a central processing unit.* M POZNIC. J. Biosci. 2009, 34 305–312.

*RNA Editing.* Bass BL, editor, 2002. Oxford University Press, Oxford, UK.

*RNA Interference (RNAi): Nuts & Bolts of RNAi Technology,* Engelke DR, editor, 2004. DNA Press, NY, USA.

*Stem cells and diabetes.* Berná G, León-Quinto T, Enseñat-Waser R, Montanya E, Martín F, Soria B. Biomed Pharmacother 2001 ; 55 : 206-12.

*Stem Cells from the Mammalian Blastocyst.* Rossant J. Stem Cells 2001;19:477-482.

*Telomeres,* deLange T and Lundblad, V, editors, 2nd ed., 2006. Cold Spring Harbor Laboratory Press, NY, USA.

*Telomeres, Telomerase, and Telomerase Inhibition: Clinical Implications for Cancer. Ahmed* A, and Tollefsbol T. J Am Geriatr Soc 51:116–122, 2003.

*Telomeres: a diagnosis at the end of the chromosomes.* de Vries BBA, Winter R, Schinzel A, van Ravenswaaij-Arts C. J Med Genet 2003;40:385–398.

*The Double Helix.* Watson JD, editor, 1968. Atheneum, NY, USA.

*The Experimenter series: Molecular Biology and Genomics,* Mulhardt C, editor, 2007. Academic Press is an imprint of Elsevier, Burlington, MA, USA.

*The Metabolic and Molecular Bases of Inherited Disease,* Scriver CR et al, editors, 8th edition, 2001. McGraw Hill, NY, USA.

*The Structures of Life.* NIH Publication No. 01-2778, 2000.

*Tissue stem cells,* Potten C, Clarke R, Wilson J, and Renehan A, editors, 2006. Taylor & Francis Group, NY, USA.

*Transcription Factors.* Locker J, editor, 2001. Academic Press, NY, USA.

*Transformation – normalizing-redifferentiation – apoptosis sequence and the role of the mitochondria in the retinoid-pancreatic cancer model: Is it obligatory? In, Cell Differentiation Research Developments (LB Ivanova, editor),* chapter XI, pp235-247, 2007. Nova Science Publishers, Inc, NY.

# Related Web-based Resources

1. http://arjournals.annualreviews.org/loi/biochem
2. http://www.ncbi.nlm.gov/
3. http://www.vlib.org/
4. www.genome.ad.jp/kegg/regulation.
5. http://arjournals.annualreviews.org/toc/biochem/76/1
6. http://www.1cro.com/mwking/
7. http://www.aw-bc.com/mathews/ch01/frames.htm
8. http://www.protocol-online.org
9. http://themedicalbiochemistrypage.org/
10. http://freemedicaljournals.com/
11. http://www.chemie.uni-greifswald.de/~biotech/Prof-Bornscheuer.html
12. http://tutor.lscf.ucsb.edu/instdev/sears/biochemistry/tw-exp/tabs-methods-frames.htm
13. http://tutor.lscf.ucsb.edu/instdev/sears/biochemistry/tw-enz/tabs-enzymes-frames.htm
14. http://www.gwu.edu/~mpb/
15. http://cshprotocols.cshlp.org/
16. http://www.molecularstation.com/protocol-links/DNA-Protocols/
17. http://www.book4doc.com/index.php/medical_books/index.11.html
18. http://bioweb.uwlax.edu/index.htm
19. http://www.expasy.ch/
20. http://alces.med.umn.edu/VGC.html
21. http://www.genome.gov/
22. http://www.rcsb.org/pdb/
23. http://www.premierbiosoft.com/netprimer.html
24. http://pymol.sourceforge.net/
25. http://www.lcmp.jussieu.fr/sincris-top/
26. http://rebase.neb.com
27. http://www.uwosh.edu/departments/llr/home.html
28. http://www.nhgri.nih.gov/About_NHGRI/Der/Elsi/

# Index

## A

## D

## E



## F

## G



## H

## I



## J



## K



## L

**M**



**N**

## O



## P

## R

## S



## T